Advances in Bricks and Blocks for Building Construction

Mohammad Arif Kamal

Architecture Section, Aligarh Muslim University, INDIA

Published by **Materials Research Forum LLC**
Millersville, PA 17551, USA

Published as part of the book series
Materials Research Foundations
Volume 108 (2021)
ISSN 2471-8890 (Print)
ISSN 2471-8904 (Online)

Print ISBN 978-1-64490-150-2
ePDF ISBN 978-1-64490-151-9

This book contains information obtained from authentic and highly regarded sources. Reasonable efforts have been made to publish reliable data and information, but the authors and publisher cannot assume responsibility for the validity of all materials or the consequences of their use. The authors and publishers have attempted to trace the copyright holders of all material reproduced in this publication and apologize to copyright holders if permission to publish in this form has not been obtained. If any copyright material has not been acknowledged, please write and let us know so we may rectify in any future reprint.

Distributed worldwide by

Materials Research Forum LLC
105 Springdale Lane
Millersville, PA 17551
USA
http://www.mrforum.com

Printed in the United States of America
10 9 8 7 6 5 4 3 2 1

Preface

Earlier, all major buildings relied on traditional red brick until the construction industry decides to change things to suit the needs and easy access to different blocks for the construction industry. Technological progress has introduced many innovations and technologies in the field of manufacturing bricks and blocks for the building construction industry. Bricks are produced in numerous classes, types, materials, and sizes which vary with region and time period, and are produced in bulk quantities. Fired bricks are one of the longest-lasting and strongest building materials, and have been used since 4000 BC. Air-dried or sun-dried bricks, also known as mudbricks, have a history older than fired bricks and have an additional ingredient of a mechanical binder such as straw, reed, and hay.

Blocks are prefabricated material mainly used to build wall masonry. Like bricks, the blocks are stacked together and joined with mortar, usually consisting of cement, sand, and water. The blocks are solid or hollow inside to allow for steel bars and mortar filling. These blocks come in a variety of dimensions and textures, from traditional smooth surfaces to fluted or rough finishes, as well as special units for corners or beams with longitudinal reinforcements. These blocks can be incorporated creatively into our design. Although the first blocks were manufactured by hand, nowadays they are produced in an automated way and thousands of blocks can be made per hour. However, by not requiring firing, each unit can be manufactured on-site by unskilled masons.

This book 'Advances in Bricks and Blocks for Building Construction' contains nine chapters that describe the different types of bricks and blocks and also introduces the recent technological advancements for bricks and blocks for the construction industry, especially in the building construction sector. This book shall be beneficial to architects, civil engineers, building scientists, academicians and construction industry professionals.

Table of Contents

CHAPTER 1

Bricks and Blocks

1. Introduction

A few years back it would be the "only brick age" for construction. Earlier, all major buildings relied on traditional red brick until the construction industry decided to change things to suit the needs and easy access to different blocks for the construction industry. People want stability and resilience, and as people begin to feel and know about environmental protection, the blocks have come at the right time. A few years ago the National Green Tribunal issued a ban on mining by making red bricks without prior naturalization. The brick kiln industry is facing a major crisis, but soon the Fly Ash Bricks or AAC Blocks industry started as people found it easier to fabricate and the simplicity of construction was also equal. Certainly, there are pros and cons of both bricks and blocks that can make us understand what to use [1].

2. Bricks

A brick is a solid unit of a building having a standard size and weight. Its history traces back thousands of years. Clay bricks are made of fired clay. The composition of clay varies over a wide range. Usually, clays are composed mainly of silica (grains of sand), alumina, lime, iron, manganese, sulfur, and phosphates, with different proportions. Clay bricks have an average density of 125 pcf. Bricks are manufactured by grinding or crushing the clay in mills and mixing it with water to make it plastic. The plastic clay is then molded, textured, dried, and finally fired. Bricks are manufactured in different colors such as dark orange, red, dark red, brown, or dark brown, depending on the fire temperature during manufacturing (Figure 1). The firing temperature for brick manufacturing varies from 900°C to 1200°C [2].

Bricks are produced in numerous classes, types, materials, and sizes which vary with region and period, and are produced in bulk quantities. Fired bricks are one of the longest-lasting and strongest building materials, sometimes referred to as artificial stone, and have been used since circa 4000 BC. Air-dried bricks, also known as mudbricks, have a history older than fired bricks and have an additional ingredient of a mechanical binder such as straw. Bricks are laid in courses and numerous patterns known as bonds, collectively known as brickwork, and may be laid in various kinds of mortar to hold the bricks together to make a durable structure [3].

Figure 1. Typical construction of a wall made up of fired clay red bricks.

3.   Different Applications of Bricks

Bricks are mainly used as a structural unit. Since the clay bricks or burnt bricks are strong, hard, durable, resistive to abrasion and fire, therefore, they are used as a structural material in different structures such as buildings, bridges, culverts, chimneys, foundations, arches, pavement (footpath, streets), and brick flooring, etc. Brick is also used as a fire-resistant material.

Bricks can be used in different colors, sizes, and orientations to get different surface designs. As an aesthetic material bricks can be used:

- In pavements
- As facing brick
- For architectural purposes

4.   Properties of bricks

4.1   Heat protection in the summer

Bricks have a high thermal mass which is what makes them absorb more heat as compared to AAC blocks. The bricks absorb heat during the day and release it at night. You get the best possible option to keep your home warm in winter and comparatively cooler in summer. Days usually require less heat as compared tonight which is where the thermal insulation properties of bricks come to play.

4.2   Eco-friendly

Those of you who commend on the environmental issues of creating red bricks forget that these bricks are made from materials that can be easily recycled and used for landfills. The same cannot be said about blocks. They are not as recyclable as bricks as they are made of concrete. All the renovation work that calls for breaking down of old structures to build new ones would only leave behind heaps of waste products that cannot be recycled.

4.3   Durability

Red traditional bricks are known to be more durable and the structures made from them are stronger than the ones made from hollow blocks. The durability of blocks is also said to last only if they are kept well maintained, unlike bricks that can stand stiff through years if the construction is done keeping in mind the proper ways of durable construction and the material used, including the quality of brick is at par with the standards of strong buildings.

4.4   Burn less easily

Bricks are nothing but highly compressed earthen materials. The high compression makes them so dense that there is very little space left for combustion to ignite and spread. They have been given the highest fire rating. They can sustain a very heavy and fully developed fire far better than a home made of blocks or any other material.

4.5   Free of maintenance

Masonry bricks are almost free of maintenance. Once constructed, they do not require as regular maintenance as block constructions do from time to time. In the long run, they do not incur any maintenance cost while block constructions need to be taken care of after few years, or else they start falling apart.

5.   Types of bricks

There are numerous types of bricks, which have different constituents and different sizes, and different properties [4]. The prominent types of bricks are summarized are followings.

5.1   Acid resistant bricks

Acid resistant bricks are specially made for chemical plants, it is made from a mixture of refractory clay, felspar and some form of silica such as white sand and flint.

5.2   Engineering bricks

Engineering bricks are those that are different than normal clay manufacturing bricks. These clay bricks are less porous and absorb less water. These bricks are very hard, therefore used for paving in construction work.

5.3   Silica bricks

These bricks have a very high percentage of silica ranging from 95 to 97%; a small amount of lime about 1 to 2% is added to serve as a binding material. Silica bricks can stand up to high temperatures up to about 2000°C; therefore these bricks are used in the construction of chimneys brick masonry.

5.4   Refractory bricks

Refractory bricks can withstand temperatures up to 1700°C. These bricks are used in the manufacture of lime flakes, stoves, and various types of furnaces that are used in the metallurgical process. These types of bricks are manufactured in brick kilns.

5.5   Sand-lime bricks

These sand lime bricks are very strong and hard bricks produced by a process that has a chemical reaction that does not need a mechanical mixture under pressure. These bricks are stronger than clay bricks, they are similar in color and texture with sharp edges, therefore these bricks are used in decorative work due to their fine texture and uniform color.

5.6   Blue bricks

These bricks are very hard and densely made from soil with 7 to 10% iron oxide; these bricks are fired at a temperature of 1250°C. Blue bricks are used for heavy engineering construction work such as bridges.

5.7   Colored bricks

This type of bricks is rarely used in India, they are used for decoration work. This type of brick has cylindrical holes throughout its thickness. They are light in weight and require less amount of soil for preparation.

5.8   Perforated bricks

These bricks are mainly used in the manufacture of a panel of bricks for lightweight structures and multi-storey frame structures.

5.9   Fire bricks

These fire bricks are used for the manufacture of stoves, furnaces, and chimneys where bricks are required to resist very high temperatures.

5.10   Hollow bricks or hollow blocks

These types of bricks are also called cavity bricks or cellular bricks. They have a wall thickness of about 20 mm to 25 mm and are prepared from special homogeneous clay. They are lighter in weight, also reducing the transmission of heat, sound, and dampness. These hollow bricks are used in the construction of brick masonry.

5.11   Fly-ash bricks

Fly ash is a fine powder thrown as a large amount of waste material at a thermal power station. The fly-ash looks like a pozzolana, it is acidic and its main components are silica, aluminum oxide, and ferrous oxide. Small quantities of fly-ash, lime, sand, and magnesium chloride are used as chemical accelerators in the process of making fly-ash bricks. Fly-ash, sand, and lime are mixed in the ratio of 80:13:7; hydraulic presses are used to make hydraulic bricks. The semi-dried bricks are cured in a steam chamber at the appropriate pressure and temperature. The fly ash bricks are superior to traditional bricks in technical specifications, compression strength, and impermeability. These bricks are lighter in weight and 10 to 15% cheaper than traditional bricks.

6.   Advantage of brick masonry

There are many advantages of brick masonry [5]. They are summarized as below:

- Brick masonry is cheaper than stone masonry.
- Bricks are of similar size.
- Brick blocks do not require any dressing.
- Bricks are light in weight. No complicated lifting equipment is required brickwork.
- There is no problem with its availability.
- They do not require transportation for long distances.
- Brick can be made with less skilled laborers.
- The bonding strength is very good and the brickwork is more durable.

7.   Disadvantage of brick masonry

There are also many disadvantages of brick masonry [5]. They are summarized as below:

- Time-consuming manufacturing process.
- Brick masonry cannot be used in a high seismic zone.
- Bricks have very low tensile strength.
- Since bricks absorb water easily, it causes fluorescence when not exposed to air.
- A rough surface of bricks can cause mold to grow if not cleaned properly.

8. Considerations for good brick masonry

The following building construction requirement for good brick masonry work should be considered [5]:

- Good brick masonry should be used in areas where sound, hard, and well-burnt, and hardened with uniform color, shape, and size.
- The bricks should be compact, homogeneous, pores-free, with cracks, defects, air bubbles, and stone lumps.
- These bricks can be soaked in water for at least two hours before use.
- In brickwork, bricks should be placed on their beds with frogs.
- The brick course should focus on horizontal and truly vertical joints.
- Brick walls should be raised evenly with a proper bond, the height of brick masonry construction should not exceed 1.5 meters in a day.
- Finished brickwork is cured for a period of 1 to 2 weeks.
- A good brick should have a fine, compact, and uniform texture.
- Good brickwork should have compressive relative strength and durability.
- It should have maximum resistance to weathering.
- Good brickwork should be fire-resistant.

9. Blocks

Blocks are prefabricated material mainly used to build wall masonry. Like bricks, the blocks are stacked together and joined with a mortar, usually consisting of cement, sand, and water. The blocks are solid or hollow from inside to allow for steel bars and mortar filling. These blocks come in a variety of dimensions and textures, from traditional smooth surfaces to fluted or rough finishes, as well as special units for corners or beams with longitudinal reinforcements. The dimensions of these blocks range from the classic approximately 19x19x39cm which is meant for structural use, to a size of approximately

19x9x39cm for partitioning walls [6]. These blocks can be incorporated creatively into our design. Although the first blocks were manufactured by hand, nowadays they are produced in an automated way and thousands of blocks can be made per hour. However, by not requiring firing, each unit can be manufactured on-site by unskilled masons.

9.1   Cheaper

Block masonry is quite cheaper compared to its brick counterpart. They are known to cost Rs. 1,500 lesser than brick masonry per cubic meter. This is not a difference to be ignored. Reports suggest that the cost of building walls from AAC blocks come out to be 17.65 percent lesser than the cost of wall made from traditional bricks. With the lesser cost of constructing these blocks, they are also easier to make which reduces the cost even more if built at the site.

9.2   Made from waste

They do not harm or deplete nature for being made. They are made from fly ash, which is nothing but a residue of thermal power plants. Whereas red bricks consume topsoil in the production and manufacturing which is like robbing nature of its precious protective layer of soil. That is the top reason why the National Green Tribunal is so much against the red bricks.

9.3   Light in weight

AAC blocks are lighter compared to red bricks which offer them more workability, flexibility, and durability. Their dry density ratio reduces the dead load on structures which makes them more efficient and suitable for modern constructions.

9.4   Strength

Concrete blocks are known to be better in enduring earthquakes which are happening quite frequently these days, hurricanes and tornadoes. Countries that promote and encourage the usage of blocks lay great emphasis on their durability to withstand such natural hazards. The way of construction also makes them further durable and stronger.

9.5   Soundproofing

For those residing in industrial areas or areas that are near busy roads, some sort of soundproofing can be a bliss. Block walls have a higher density as compared to brick constructions and hence they offer more soundproofing. Their efficient acoustic insulation is a big help if your home is constantly surrounded by noise that could keep you from getting sound sleep. You can think of the advantage of soundproofing if you ever resided close to railways or airports.

9.6   Space saving

Builders and contractors are mostly recommending concrete blocks because they save quite some space. The width is less and durability doesn't decrease which adds to the space required in building walls. The usual 9-inch walls of the traditional bricks are getting replaced for good, especially since there is a lot of fight for space in big cities.

10.   Advantages of concrete building blocks

There are many advantages of concrete blocks [7]. They are summarized as below:

10.1   Strength and durability

Concrete blocks are strong, and they can stand the test of time. According to the World Business Council for Sustainable Development's Cement Sustainability Initiative, concrete gets stronger over time and is not weakened by moisture, pests, or mold.

10.2   Fire resistance

Where a wooden structure will quickly succumb to flames, a concrete block building can resist them and limit fire's ability to spread quickly.

10.3   Locally manufactured

Concrete blocks are almost always manufactured and sold locally due to the high costs associated with transporting heavy materials.

10.4   Insulating properties

Concrete blocks are known for their insulating properties. Concrete blocks excel at keeping outside temperatures from entering the home, often resulting in lower energy costs. These insulating properties vary by manufacturer and depend on the block's density. Hollow concrete blocks can be filled with insulation during construction if desired.

10.5   Custom colors and finishes

Today, concrete building blocks can be custom ordered to best meet your project's requirements. Not only can you specify the color and finish, but concrete blocks can also be made in various shapes and sizes.

10.6 Affordable and sustainable

Oftentimes, environmentally friendly choices come with a hefty price tag. Concrete blocks remain affordable, and they are increasingly being embraced as an environmentally-friendly choice. The raw materials used to produce concrete blocks are sustainable and recyclable. Producing them uses less energy than it does to produce comparable building materials, and the CO_2 emissions associated with concrete block production are considered relatively low compared to those of similar building materials.

11. Difference between red bricks and solid concrete blocks

Red bricks are one of the oldest and extensively used building material that is primarily made from clay. Solid concrete blocks, on the other hand, are precast concrete blocks manufactured from cement and fine aggregates. The important difference between red bricks and solid concrete blocks are tabulated below [7].

S. No	Parameter	Red Bricks	Solid Concrete Blocks
1	Raw Materials	Red bricks use the following raw materials: LimeClay or Alumina, Sand, Iron Oxide, Magnesia. The sand used for red brick manufacture is mostly obtained locally.	Solid Concrete blocks use the following raw materials: Ordinary Portland Cement, Sand, Gravel, Water. In certain situations, fly ash can be used instead of fine sand.
2	Properties	Red bricks are available in modular sizes of 190 x 90 x 90mm and 190 x 90 x 40mm. And also, in Non-modular sizes of 230 x 110 x 70mm and 230 x 110 x 30mm	Standard sizes of solid concrete blocks are of length 400, 500, or 600mm and height of 200, 100mm and the width from 50, 75, 100, 150, 200, 250, or 300mm. The dimension differs from manufacturer to manufacturer.
3	Compressive Strength	The compressive strength varies from one class to another and hence lies in the range between 3.5 to 35N/mm².	The compressive strength of a solid concrete block varies based on the grade of cement used. Its compressive strength varies from 4 to 5N/mm².
4	Dry Density	The dry density also varies depending on the class of brick. It normally ranges from 1600 to 1920 kg/m³	The dry density of the solid concrete block is depended on the grade of the block. This ranges from 1800 to 2500 kg/m³

5	Water Absorption	The water absorption of red bricks is recommended to be less than 20% of its weight	Solid concrete blocks must not have a water absorption value not greater than 10% of their weight
6	Thermal Conductivity	The thermal conductivity of red bricks is advised to have a value in the range of 0.6 to 1W/mK.	Solid concrete blocks usually have a thermal conductivity in the range of 0.7 to 1.28W/mK.
7	Environmental Impact	Red bricks utilize naturally available clay. This production hence depletes the top fertile soil. Red bricks also emit more carbon dioxide during their manufacture.	The amount of carbon dioxide emitted during the manufacture of solid concrete blocks is less.
8	Mortar Consumption	The mortar consumed by red bricks is high due to its irregular surface.	Solid concrete blocks have flat and even surfaces that hence demand less mortar compared to redbricks.
9	Water Usage	Curing requires more water	Solid concrete blocks require 7 to 14 days of curing which demands a high amount of water compared to red bricks.
10	Cost	Red bricks alone are cheap. But overall cost including the cost of mortar and construction is high as it demands more mortar.	The solid concrete blocks cost high as individual pieces. It consumes less mortar. It has the advantage that the same wall area can be constructed with less number of solid concrete blocks than red bricks.
11	Uses	Red bricks can be used as a structural material for the construction of structures like buildings, foundations, arches, pavement, and bridges. These can be also used for aesthetic purposes like landscaping, facing works, and many other architectural purposes.	Solid concrete blocks are employed in construction to act both as load-bearing and non-load bearing in walls, panel walls, and partition walls. This can also be used as backing for piers, retaining walls, other facing materials, chimneys, fireplaces, and garden walls, etc.

Conclusion

The Brick v/s block debate is a very close one and anyone who is about to get construction started to have will have to deal with the comparing of pros and cons of both. However, a lot depends on the place, budget, and situation of the builder. For some, brick masonry constructions are necessary because they ultimately want lesser maintenance even when the cost of building is higher. Some people, on the other side, would prefer less cost of construction and regular renovations to keep the blocks well maintained. The making of red bricks is not so eco-friendly it seems but the same can be said about the blocks not being up for recycle. Redbrick kilns remain under the scrutiny of the government and National Green Tribunal while the concrete block industry is on the rise at many places.

References

[1]   Information   on   https://www.careerride.com/view/bricks-or-blocks-which-are-better-for-construction-26826.aspx

[2]   Information on https://www.aboutcivil.org/bricks-advantages-disadvantages-uses.html

[3]   Information on https://en.wikipedia.org/wiki/Brick

[4]   Information on https://constructionor.com/brick-masonry/

[5]   Information on https://theconstructor.org/building/difference-red-bricks-solid-concrete-blocks/37090/

[6]   Information   on   https://www.careerride.com/view/bricks-or-blocks-which-are-better-for-construction-26826.aspx

[7]   Information   on   https://www.asanduff.com/advantages-of-concrete-building-blocks/

CHAPTER 2

Fly Ash Bricks

1. Introduction

Fly ash is a siliceous material with the association of amorphous/glassy mass. The constituent particles of fly ash react with lime at elevated temperature and pressure of steam to form calcium silicate-hydrate and calcium aluminate hydrate. The nature and extent of formation of calcium silicate hydrate depend upon many factors like physicochemical characteristics of raw materials, the molar ratio of calcium oxide and silica, and curing conditions. Fly ash brick (FAB) is a building material, specifically masonry units, containing class C or class F fly ash and water. Compressed at 28 MPa (272 atm) and cured for 24 hours in a 66 °C steam bath, then toughened with an air-entrainment agent, the bricks can last for more than 100 freeze-thaw cycles. Owing to the high concentration of calcium oxide in class C fly ash, the brick is described as "self-cementing". The manufacturing method saves energy, reduces mercury pollution in the environment, and often costs 20% less than traditional clay brick manufacturing.

The use of cement, fly ash, and phosphogypsum is made as an alternative for burnt clay bricks. This is a new technology that works with the strength of fly ash, lime and gypsum chemistry. The slow chemistry of fly ash and input is maneuvered by tapping a hydrous mineral phase to its threshold limits through a sufficient limit of gypsum. Therefore, it does not require a heavy-duty press or autoclave, which is otherwise required in the case of only fly ash and lime. The process requires open-air drying and curing and does not require combustion of any fossil fuel. To gain early strength autoclaving with steam curing can be adopted. The ingredients of the units such as bricks and blocks, fly ash, lime (from OPC), and gypsum are well-known minerals that are widely used in the industries. All these minerals are available in the form of wastes and by-products. If lime is not available in adequate quantity, Ordinary Portland Cement (OPC) can be used as the replacement of lime producing a good quality of bricks and blocks. This technology is proved to be environmentally safe and sound.

The coal dust has historically been collected as a waste product from homes and industry. During the nineteenth century, coal ash was taken by 'scavengers' and delivered to local brickworks, where the ash would be mixed with clay. The income from the sale of ash would normally pay for the collection of waste [1]. The clay is typically entrapped during the formation of coal. When coal is burnt, the incombustible clay particles are left behind as ash. Ingrate boilers, incombustible ash agglomerates as cinders through prolonged residential time. Nowadays, pulverized coal technology is preferred due to its improved energy efficiency. In this case, the ground clay escapes along with flue gases, settling as ash in bag filters or electrostatic precipitators (ESPs). This gives rise to the name 'fly ash' [2]. Figure 1 shows fly ash bricks with a frog.

Figure 1.   Fly Ash bricks with frogs.

2.   Ancient lime-pozzolanic chemistry

When fly ash is blended with lime, the reactive silicates and aluminates form into hydrated mineralogy as follows:

$$Ca(OH)_2 + SiO_2 + H_2O \quad \rightarrow \quad C_3S_2H_3$$

$$Ca(OH)_2 + Al_2O_3 + H_2O \quad \rightarrow \quad C\text{-}A\text{-}H$$

In the above mineralogy, calcium silicate hydrates do render strength but take a longer time for formation. Though calcium aluminate hydrates do form relatively at early ages, the mineralogy renders with feeble strength.

To overcome these rheological shortcomings for commercial production there needs to be a solution. That is what is Fal-G. Erstwhile technologies for manufacturing fly ash bricks advocated the use of the heavy-duty press in association with autoclave, both to overcome the disadvantage of slow reactions between fly ash and lime. The capital-intensive plant and machinery on one hand and the energy-intensive process operations on the other have their cascading effect on the ultimate cost of the product, thus making the same cost-prohibitive. This scenario prevented fly ash utilization in the brick segment, more so in the second and third-world countries due to low marketing prices for walling material [3].

3.   Types of fly ash bricks

Keeping in view the raw material, fly ash bricks can be mainly grouped into two categories:

- Fly ash bricks using cement as a binder

Raw materials include: Fly ash, cement, and sand

- Fly ash brick using lime as a binder

Raw materials include:; Fly ash, lime, gypsum, and sand; also known as Fal-G.

Or

Fly ash, Lime, and Sand.

Fly ash is a fine residue obtained from thermal power stations using ground or powdered coal as boiler fuel. It can be utilized in various forms as a building material. The thermal power stations in the country throw large quantities of fly ash which goes as waste but which could be effectively used as partial replacement of cement.

The physical and chemical properties of fly ash obtained from different thermal power stations vary widely. There will be a considerable difference in these properties of fly ash even if the fly ash is obtained from the same power station depending upon the coal used by the power stations.

However, the range of variation permissible, as regards the chemical property is concerned, is so wide that most of the fly ash obtained from the power stations is suitable for use as pozzolanic material. IS: 3812 (Part II) gives the permissible range of chemicals and their concentration.

As regards the physical properties BIS prescribes a fineness corresponding to a specific surface area of 3200 cm^2/gm but the fly ash normally obtained is found to have a specific surface area of 2500 to 6000 cm^2/gm. The BIS prescribes a minimum pozzolanic activity of 200m^2 per kg and the grade I and grade II fly ash normally obtained fully satisfies this stipulation.

4. The FaL-G technology

The chemistry of fly ash-lime has its antecedents in civil engineering practices for the last two millennia. Most of the Roman constructions have been executed with pozzolana-lime mixes. The formulations in those days were based on empirical judgments in the absence of chemical engineering explanations. But, physical behavior was judged as the basis to decide the quality of products. After the advent of Ordinary Portland Cement (OPC), the concern to optimize pozzolana-lime chemistry was not felt much. Since the fall of the 60s, whilst the confidence in OPC as the durable media has begun waning out, pozzolana-lime chemistry has again gained importance, OPC being the source of lime. This is how PPC has emerged as a cement blend. In this background, FaL-G has made its presence felt, proving the factor of innovation in its composition through the optimized role of calcium alumino sulphate mineralogy [3].

FaL-G is an innovative cementitious binder that can offer structural concrete with sound microstructure properties. In simple terms, it is an extension of lime-pozzolanic binder by adding the specified quantity of gypsum to tap the potential of calcium sulpho aluminate hydrates as follows:

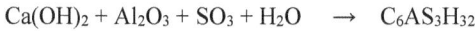

$$Ca(OH)_2 + Al_2O_3 + SO_3 + H_2O \quad \rightarrow \quad C_6AS_3H_{32}$$

5.   Specifications and characteristics of fly ash bricks

5.1   General requirements

Visually the bricks shall be sound, compact, and uniform in shape. The bricks shall be free from visible cracks, warpage, and organic matter.

The bricks shall be solid and with or without the frog. The frog can be 10 to 20mm deep on one of its faces.

5.2   Dimensions and tolerances

The standard modular sizes of fly ash bricks are given in Table 1.

Table 1.   The modular sizes of Fly-Ash bricks

Length (L) mm	Width (W) mm	Height (H) mm
190	90	90
190	90	40

The non-modular sizes of the bricks which are produced are given in Table 2.

Table 2.   The non-modular sizes of Fly-Ash bricks

Length (L) mm	Width (W) mm	Height (H) mm
230	110	70
230	110	30

5.3   Tolerances

The dimensions of bricks when tested shall have tolerance in length by ± 3 mm and that in breadth & height by ± 2 mm.

5.4   Classification

The different classes of fly ash lime bricks, depending upon their average compressive strength are given in Table 3.

Table 3. Different classes of Fly ash bricks

Class Designation	Average Compressive Strength Kg/cm^2 (approximate)
30	300
25	250
20	200
17.5	175
15	150
12.5	125
10	100
7.5	75
5	50
3.5	35

5.5   Physical characteristics

Compressive strength

The minimum average compressive strength of fly ash lime bricks shall not be less than the one specified for each class in IS: 3495(Part-I) -1976. The compressive strength of any individual brick shall not fall below the minimum average compressive strength specified for the corresponding class of bricks by more than 20 percent.

Drying shrinkage

The average drying shrinkage of the bricks, when tested by the method described in, IS: 4139-1989, being the average of three units, shall not exceed 0.15 percent.

Advances in Bricks and Blocks for Building Construction   Materials Research Forum LLC
Materials Research Foundations **108** (2021)   https://doi.org/10.21741/9781644901519

Efflorescence test

The bricks when tested by the procedure laid down in IS:3495 (Part-3)-1976, shall have the rating of efflorescence not more than 'moderate' up to class 10 and 'slight' for higher classes.

Water absorption

The bricks, when tested by the procedure laid down in IS:3495 (Part-2)-1976, after immersion in cold water for 24 hours, shall have average water absorption not more than 20 percent by mass up to class 12.5 and 15 percent by mass for higher classes.

6.   Manufacturing process

For the manufacturing of Bricks, the composition of materials is as below:-

6.1   For fly ash – lime bricks

Fly ash   65 – 70%

Sand   15 – 20%

Lime   5 – 10%

Gypsum   3.5 – 6.5%

6.2   For fly ash – cement bricks

Fly ash   50 – 70%

Sand   20 – 30%

Cement   8 – 12%

Another composition is also feasible wherein both the binders lime and cement can be used.

6.3   For fly ash – cement – lime bricks

Fly ash   50 – 70%

Sand   20 – 30%

Lime   5 – 7%

Cement   2 – 5%

A Typical Process flow diagram for the manufacture of Fly ash Sand Lime Bricks is given below in Figure 2 and a semi-automatic and automatic machine developed by BMTPC, New Deli, India is shown in Figure 3 and Figure 4 respectively.

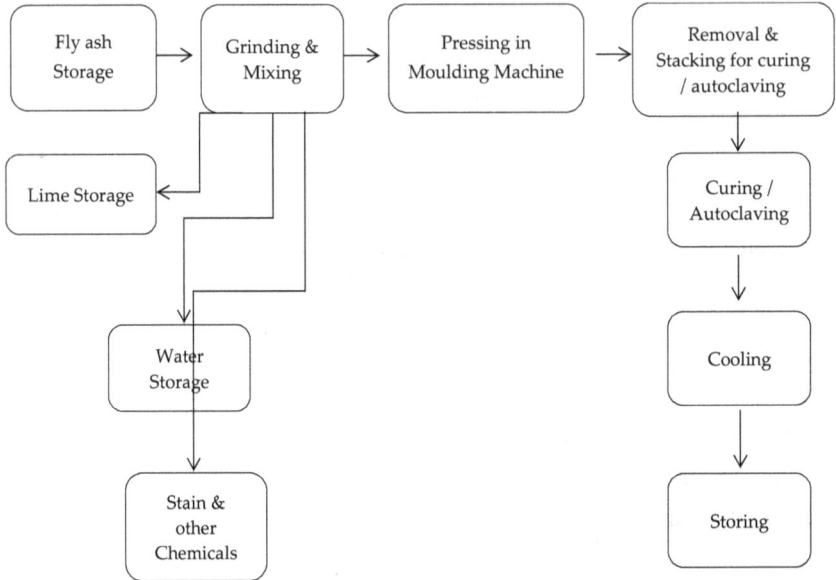

Figure 2. Schematic diagram of manufacturing of fly ash-lime-gypsum bricks (FaL-G bricks) showing the process as well as the machines.

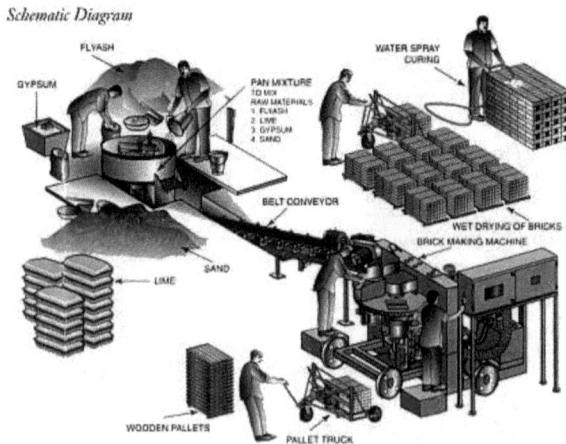

Figure 3. A semiautomatic machine for manufacturing fly ash bricks.

Figure 4. An automatic fly ash brick making machine developed by BMTPC, India.

7.   Properties of fly ash affecting the strength of fly ash bricks

The following properties of fly ash affect the strength and look of fly ash bricks:

- Loss on Ignition (LOI): fly ash loses weight when it burns at about 1000°C due to the presence of carbon and water. The weight loss that happens due to carbon combustion and moisture evaporation is called 'Loss on Ignition (LOI)'. This is expressed as a percentage. The lower the loss of Ignition, the better the fly ash. As per BIS, it should not be more than 5%.

- Fineness: fine fly ash has more surface area available to react with lime. This increases pozzolanic activity, which contributes to the strength of fly ash bricks. As per BIS, it should not be more than 320 m^2/kg.

- Calcium (CaO) content: the pozzolanic reactivity of fly ash is more in high calcium fly ash. The greater the pozzolanic activity leads to higher the strength of fly ash brick. As per ASTM C618, fly ash is classified into two types: Class C contains more than 10% lime and Class F fly ash contains less than 10% lime.

Based on boiler operations, fly ash can be additionally classified as LT (low temperature) and HT (high temperature). LT fly ash containing amorphous phases is generated where boiler temperature is not more than 800°C, whereas HT fly ash containing glassy reactive phases is generated at more than 1000°C in super thermal plants. LT fly ash reacts well with lime whereas HT fly ash reacts well with OPC [2].

8.   Benefits of fly ash bricks for the environment

The increase in greenhouse gases, out of which CO_2 is one of the major constituents, increases global warming year after year, causing drought and floods. The total CO_2 emissions globally account for 24,960 million tons at 1990 levels. The cement and building materials industry is one of the major contributors. The CO_2 emission is about ninety million tons out of cement and forty-nine million tons out of clay bricks production in India. As per the ongoing practices in India, each million clay bricks consume about 200 tons of coal (or any other fuel with an equal quantity of thermal values) and emit around 270 tons of CO_2. Fly ash bricks production in energy-free route saves the emissions completely, befitting the project to qualify under Clean Development Mechanism (CDM), as envisaged by the Kyoto Protocol towards the welfare of Mother Earth [3]. The FaL-G is one of the rare technologies to serve all the indicators of Sustainable Development. They are summarized in Table 4.

Table 4. Benefits of Fly ash Bricks for the Environment

Sustainable Development Indicators	Redeeming features and tangible results in the field
Conservation of Natural Resources	Saves precious topsoil otherwise used for clay bricks.
Conservation of thermal energy and fossil fuels.	Saves coal as no sintering is involved as practiced for clay bricks.
Environment friendly	Used the Industrial by-products such as fly ash, lime, and chemical gypsum, avoiding resultant pollution.
Employment generation	Provides yearlong employment, unlike clay bricks where the production activity is seasonal (November to May), with more scope in rural areas.
Appropriate and eco-friendly technology	Renders more durable products for housing and infrastructure applications with longer service life, conserving mineral and fiscal resources to the Nation.

9.   Durability of fly ash bricks

It has been observed that there is an increase in Cold Crushing Strength (CCS) of fly ash bricks concluding that buildings made with fly ash bricks will grow stronger with time. Efflorescence is another cause of worry to the buildings constructed using burnt clay bricks particularly in the areas with high salinity in the underground water or the atmosphere. Fly ash brick is impervious to efflorescence being free of soluble salts. The absence of salts in fly ash bricks protects it completely from salinity. It is thus ideal for constructions on seafronts or where underground water is high in salinity where burnt clay bricks cannot endure. However, the quality of fly ash bricks is of utmost importance to ascertain durability that is why the fly ash bricks are usually not recommended in foundations.

10. Advantages of fly ash bricks

- It reduces the dead load on structures due to lightweight (2.6 kg, dimension: 230 mm X 110 mm X 70 mm.

- The same number of bricks will cover more area than clay bricks

- Fly ash bricks have high fire Insulation

- Due to high strength, practically no breakage during transport and use.

- Due to the uniform size of bricks mortar required for joints and plaster reduces almost by 50%

- Due to lower water penetration seepage of water through bricks is considerably reduced.

- Gypsum plaster can be directly applied to these bricks without a backing coat of lime plaster.

- These bricks do not require soaking in water for 24 hours. The sprinkling of water before use is enough.

- Fly ash bricks are much superior to clay bricks available in local areas and as such can prove to be a boon in the building construction field.

11. Disadvantages and limitations of fly ash bricks

- Depending on the mixture mechanical strength can be less. This can be partially rectified by adding marble waste or mortar between blocks.

- Large size can have more breakages depending on the mix of materials.

- It has high thermal conductivity. Extra insulation is required in colder regions.

- The longevity of fly ash bricks in the sub-soil is yet to be established. Thus its use in foundations is not recommended at present.

- The quality of clay bricks generally available in the market is poor. The edges are broken during stacking and transportation. These bricks also absorb more water. The quality of fly ash bricks depends upon the quality of lime, proper mixing of ingredients, curing period. Incidentally, the manufacturers also adopt cut-short methods, which degrade the quality of fly ash bricks. The manufacturers are small production units and cannot stack the bricks and hold them for a longer period to allow the full curing process to complete.

- In order, to produce quality fly ash bricks in terms of strength, shape, and size with sharp edges, it is recommended to use 5% to 7% of cement by weight. Mixing of more than 7% cement creates shrinkage cracks. However, proper dry mixing of all ingredients helps in maintaining the uniform strength of bricks.

- The use of well-made fly ash bricks in exposed brickworks has shown good results without erosion and with time. The bonds in the masonry gain strength with time as a result of the slow reaction of cement/lime with fly ash.

12. Comparison of fly ash bricks with clay bricks

Fly ash bricks are hi-tech well-improved quality bricks used for the construction of brick masonry structures. They are used as a replacement for normal clay bricks and have better properties than them. Fly ash bricks competitive in comparison to conventional clay bricks and provide enormous indirect benefits. The utilization of fly ash bricks results in the conservation of natural resources as well as the protection of the environment [3]. The comparison of fly ash bricks with clay bricks is shown in Table 5. An example of modular fly ash bricks in a load-bearing four storied housing at Baprola, Delhi is shown in Figure 5.

Table 5. Comparison of fly ash bricks with clay bricks

Properties	Red Bricks/Clay Bricks	Fly Ash Bricks	Remarks
Density	1600-1750 kg/m^3	1700-1850 kg/m^3	Higher load-bearing
Compressive strength	30-35 kg/cm^2	90-100 kg/cm^2	Higher load bearing
Absorption	15-25%	10-14%	Less dampness
Dimensional stability	Very low tolerance	High tolerance	Saving in mortar up to 25%
Wastage during transit	Up to 10%	Less than 2%	Saving in cost up to 8%
Cement Mortar	Cement mortar required is 23–25%	Cement mortar required is 8–10%	Saving in mortar due to even shape
Plastering	Thickness varies on both sides of the wall	Even on both sides	Saving in plaster up to 15%.

Figure 5. Modular fly ash bricks in a 4-storied housing at Baprola, Delhi, India.

Conclusion

Fly ash is no more just waste. It is the potential resource for the production of brick, cement, and concrete. It is the most versatile material to serve the engineering properties right from brick to concrete without sacrificing the economics. It would be a default of the highest magnitude to ignore the products such as fly ash that have multifarious contributions to the benefit of every stakeholder. Fly ash was subjected to such abuse during the '70s, in the absence of knowledge and awareness. It is time that the availability and distribution of this potential material are channelized supported by technical bulletins and education. Improper use of any material is detrimental to its performance.

References

[1]   E. Chadwick, Report to Her Majesty's Principal Secretary of State for the Home Department, from the Poor Law Commissioners, on an Inquiry into the Sanitary Condition of the Labouring Population of Great Britain. Clowes for HMSO. (1842) p. 53.

[2]   N Bhanumathidas and N Kalidas, Fly ash for Sustainable Development, Ark Communications, India, 2002.

[3]   N Bhanumathidas and N Kalidas, FaL-G : The Technology from Brick to Cement Concrete, Course Material, Institute for Solid Waste Research & Ecological Balance (INSWAREB), Visakhapatnam, India, 2007.

[4]   Information on https://theconstructor.org/building/fly-ash-bricks/5330/

CHAPTER 3

Calcium Silicate Bricks

1.   Introduction

Calcium silicate bricks are made of sand and lime and are popularly known as sand-lime bricks. These bricks are used for several purposes in construction industries such as ornamental works in buildings, masonry works, etc. Sand lime bricks are popularly used in European countries, Australia, and African countries. In India, these bricks are widely used in the Kerala state and their usage is regularly growing. Calcium silicate bricks are suitable for use in both external and protected internal walling. They are available as facing bricks or as commons. As for clay bricks, the bricks are available in a solid or a frogged unit and are made to a standard size of 216 x 102.5 x 65 mm. The method of manufacture together with inherent properties of the mixed raw materials produces a brick with fine dimensional tolerances and good clean arises. Figure 1 shows the colored Calcium Silicate or Sand Lime Bricks.

Figure 1. The colored calcium silicate or sand lime bricks.

2.   Materials used for sand lime bricks

The raw materials used in the manufacture are a very fine siliceous aggregate, high calcium lime, and water. Inert and stable pigments are normally added to give the required color. The materials are first mixed in the required proportions and are mechanically pressed under considerable pressure into molds. They are then cured in high-pressure steam autoclaves for several hours which results in the combination of the lime with part of the siliceous aggregate to produce a hydrous calcium silicate (tobermorite) which forms the binding medium in the finished brick. The materials listed below are used for the production of calcium silicate bricks [1]:

2.1   Sand

Calcium silicate bricks contain a high amount of sand is about 88 – 92%. It means the properties of these bricks depend upon the characteristics of sand used. Sand has two distinct functions to perform in the manufacture of sand-lime brick, which require different properties of the material. Part of it must enter into a chemical combination with the lime to form the calcium-silicate bonding material. The rest of the sand grains constitute the aggregate which is bound together and which forms the main body of the brick. It is necessary, therefore, that that part of the sand which is expected to combine with the lime shall be in as fine a state of division as possible. If this does not occur naturally, some of the sand must be ground until it is of sufficient fineness. A good practical rule is that about 15% of the sand must pass a 100-mesh screen. The remaining 85 % of the sand is intended to form the inert filler or main body of the brick. The sand used shall be well-graded and should not contain any impurities like organic matter, soluble salts, etc. the finely divided clay may be present but it is up to 4% only which helps the brick in pressing and provides smoother texture.

2.2   Lime

Lime content in calcium silicate bricks varies from 8 to 12%. The lime used shall be of good quality and high calcium lime. Although the proportion of lime used in making sand-lime brick is relatively small, its quality is of paramount importance. The lime must be perfectly hydrated before the bricks are pressed. Otherwise, it will expand during the steam treatment and produce internal strains which are frequently sufficient to disrupt the brick. The lime must also be sufficiently caustic to enter readily into combination with the sand

2.3   Water

Clean water should be used for preparing calcium silicate bricks. Seawater or water containing soluble salts or organic matter of more than 0.25% is not suitable.

2.4   Pigment

Pigments are generally used to give color to the bricks. They are added to the sand and lime while mixing. The total weight of brick contains 0.2 to 3 % of pigment quantity. Different pigments used to get different colors are tabulated in Table 1.

Table 1. The different constituent materials used for sand lime bricks

Pigment	Color
Carbon black	Black, grey
Iron oxide	Red, brown
Chromium oxide	Green
Ochre	Yellow

3. History of calcium silicate bricks

The process for making brick of sand was discovered and patented by Dr. William Michaelis in 1880. Peppel indeed recognizes sand-lime brick made by other processes, and numerous statements may be found in the literature concerning a "sand-lime brick" made in New Jersey about 1860. This material was a mortared brick, being made of ordinary lime mortar, molded into shape, and permitted to set. So far as can be learned, none of these earlier processes has proved a commercial success. All sand-lime bricks now marketed in this country are made according to the basic principles covered by the original Michaelis patent. Dr. Michaelis permitted his patent to lapse without exploitation, almost immediately thereafter several modifications appeared.

4.   Manufacturing of calcium silicate bricks

Calcium silicate (sand lime and flint lime) bricks are manufactured by mixing lime, sand, and/or crushed silicaceous or flint stone together, with enough water to allow the mixture to be molded under high pressure. The bricks are then steam autoclaved so that the lime reacts with the silica to form hydrated calcium silicates. Pigments can be added during the mixing stage. In their natural state, calcium silicate bricks are white to a creamy off-white color, but the addition of ochres (buff or cream colors), iron oxides (pink, red, brown, or black), or chrome oxide (green) can enable a very wide variety of colors to be produced [2].

4.1   Proportioning of materials

The relative proportions of sand and lime which are used in the manufacture of sand-lime brick are of course very important in determining the properties of the finished product. In the first step, suitable proportions of sand, lime, and pigment are taken and mixed thoroughly with 3 to 5 % of water. Then paste with moldable density obtains. The mixture is molded into bricks using a rotary table press which uses mechanical pressure to press the bricks. The pressure of pressing varies from 31.5 to 63 N/mm^2.

It will be remembered that Michaelis's patent called for "from 10 to 40 parts of calcium hydroxide to 100 parts of sand. "The pure calcium silicate is a gelatinous material which shrinks on drying like lime paste. To obtain a lime mortar of maximum strength, it is necessary to add enough sand to the paste so that the internal strains caused by this shrinking will be largely overcome. The hydrate is therefore 25 percent by volume of the sand. Assuming that sand weighs 100 pounds per cubic foot and hydrated lime weighs 40, the hydrate is about 10 percent by weight of the sand. If the hydrate carries 24 percent water the above proportions are equivalent to 7.6 parts of quicklime to 100 parts of sand by weight.

4.2   Mixing of materials

Probably the most important step in the manufacture of sand-lime brick is the mixing of lime and sand. This operation is usually the determining factor in the quality of the brick. Every precaution should be taken that the mixing is thorough and efficient so that the best results be obtained from the raw materials at hand. The lime and fine sand should be in intimate contact with each other so that the chemical reaction between them can readily take place. The coarse sand should be evenly distributed throughout the mass so that the proportion of voids shall be a minimum.

4.3   Compaction of materials

The mixture of lime and sand is now ready to be pressed into the form of a brick. Pressing serves not only to give the brick its final size and shape but performs several more important functions. By bringing the sand and lime into very intimate contact with each other, the chemical combination between them can be facilitated. The compression of the material necessarily decreases the proportion of voids and therefore produces a less porous brick. The final strength of the brick has been found to depend, to some extent, on the pressure exerted in molding it. A sand-lime brick is not subject to any other change of shape or size after it leaves the press.

4.4   Hardening of bricks

The bricks are picked by hand from the press table and piled upon iron trucks. This operation requires considerable care, for at this stage of the process the bricks are very tender and easily crushed in the hand. The trucks are designed to carry about 1100 bricks each. As soon as a truck is loaded it is pushed by hand on tracks into the hardening cylinder. The bricks are subjected to a steam treatment and can be made to almost any size or shape.

In the final stage, bricks are placed in an autoclave. An autoclave is nothing but a steel cylinder with tightly sealed ends. The diameter of the autoclave is about 2m and the length is about 20m. This is a cylindrical shell of open-hearth steel, 70 feet long by 6 feet in diameter, built up of plates five-eighths of an inch thick, riveted together. After placing bricks in this closed chamber saturated steam pressure is released at about 0.85 to 1.6 N/mm^2. The temperature inside the chamber is raised and the reaction process begins. The shell is set up horizontally and tracks for the cars laid on the bottom of it. The front end is used as a cover and is held in position by 50 nickel-steel bolts 1 inch in diameter. After erection, the cylinder is tested to a pressure of 225 pounds per square inch. Thirty-three such cylinders will hold 20 trucks, carrying 22,000 bricks. The cover of this cylinder is designed to be lifted off using a small chain block. The press is operated during the day until the cylinder is filled, and the steam treatment takes place during the night. Since it requires about 3 hours to bring the cylinder up to maximum pressure and about 1 hour to blow off the steam before the cover can be removed, the duration of curing is limited to about 10 hours. Under these circumstances, a steam pressure of about 120 pounds per square inch has been found satisfactory. Silica content in sand and calcium content in lime reacts and forms a crystal-like compound called calcium hypo silicate. This process is done for 6 to 12 hours. Finally, the obtained bricks are transported to the workplace.

Figure 2. The 20m long steel cylinder act as an autoclave.

4.5   Stacking and handling of bricks

When the trucks are taken out of the hardening cylinder, they should immediately be taken to the loading platform, where the bricks are transferred to freight cars for shipment. The bricks are generally handled by one of three methods:

i.   Passing them by hand from one man to another;

ii. The hand carrier, which is built on the principle of a pair of tongs and is capable of picking up 8 or 10 bricks

iii. The gravity carrier is a chute set on small iron legs arranged to give it a slight inclination. The bed of the chute is made up of small wooden rollers, set close together, which are free to turn as the bricks slide over them.

Figure 3. The stacking of calcium silicate bricks.

5.   Properties of sand lime bricks

The following properties of sand-lime bricks are summarized as follows [3]:

- They have a very smooth and uniform finish, it has an attractive appearance.
- They are dense, strong, and tough.
- They are porous therefore free from indigestion.
- Also, uniform in size, shape, finish, and no plastering is required.
- When required, the amount of plaster is significantly less.
- Essential materials are quite common in the event can be used as an alternative to clay bricks

6.   Advantages of calcium silicate bricks

There are many advantages of calcium silicate bricks when used in masonry construction, and they are:

- Mortar required for providing plaster on calcium silicate bricks is very less.
- The color and texture of these bricks are uniforms.
- The compressive strength of sand-lime bricks is about $10N/mm^2$. So, they are well suitable for multi-storied buildings.
- For constructions in clay soils, these bricks are more preferable.
- The trouble of efflorescence does not arise in the case of sand-lime bricks.
- Not only bricks, blocks, and tiles can also be made using calcium silicate.
- Sand lime bricks provide more comfort and accessibility for architects to attain desired shape and designs.
- These bricks have accurate shape and size with straight edges.
- Solar heat effect is reduced on exposed walls made of calcium silicate bricks.
- Colored sand-lime bricks do not need any finish to the wall, so, the cost reduces.
- These bricks have great fire resistance and water repellant properties.
- Calcium silicate brick walls resist noise from outside.
- The cost of construction gets reduced by about 40% of the total cost due to the following factors.
- Wastage of calcium silicate products is very less.
- Less quantity of mortar is needed.
- The thickness of the wall can be reduced when constructed using these bricks because of high compressive strength.

7.   Disadvantages of calcium silicate bricks

In some conditions, calcium silicate bricks are not suitable and their disadvantages are:

- If the clay is available in plenty, clay bricks are more economical than calcium silicate bricks.
- These are not suitable for laying the foundation, because they cannot provide resistance to water for a longer period.
- They cannot resist fire for longer periods so, they are not suitable for building furnaces, etc.
- The abrasion resistance of these bricks is much less so, they cannot be used as paving materials.

8. Problems associated with calcium silicate bricks

Some problems are associated with calcium silicate bricks [4]. They are summarized as below:

- Thermal movement is likely to be about 1.5 times that of clay brickwork. Calcium silicate brickwork, unlike clay, usually undergoes an initial irreversible shrinkage on laying (clay brickwork tends to expand) but so long as the propensity for movement is understood and catered for in the design, there is no reason why the brickwork should not perform adequately. Often this factor is not catered for in the design and this results in widespread cracking.

- Calcium silicate bricks should not be used in solid work with clay facings or backings, this is because of the propensity of the bricks to shrink in contrast with the expansion of clay brickwork. If solid walling is to be contemplated, backings of concrete bricks or blocks should be used, as these have similar movement characteristics to calcium silicate bricks. We often see an inappropriate choice of walling material for the inner leaf and this sets up opposing forces due to differential expansion, again resulting in widespread cracking.

- General construction detailing is often not attended to, particularly about providing sufficient flexibility in the wall ties to permit the differential movements and allowing for discontinuity around cavity closers to prevent cracking.

- The requirement for inbuilt slip planes is often not attended to. Internally, walls of calcium silicate brickwork need to be bedded on a damp proof course to act as a slip plane and so facilitate longitudinal movements to occur – this would be equally necessary at upper floor levels, a detail that had been missed at this scheme.

- Movement control in walling is not the only issue – also consider building elements that could provide a restraining influence. For example, concrete columns or walls cast up against bricks should be avoided unless a slip membrane can be provided. – as should any form of construction that will prevent free movement. At this scheme pointing of the movement joints and DPC, both provide this restraining influence.

- It is not unusual to see some form of displacement with calcium silicate bricks due to thermal expansion, for example, brickwork sliding off a damp proof course, cracking at corners, or evident disruption. By contrast, shrinkage cracking does not generally produce these manifestations.

Conclusion

Calcium silicate bricks are often given bad press due to the issues highlighted here; however, it should be said that they are an excellent building material so long as the construction detailing required dealing with shrinkage or expansion is understood. Unfortunately, more often than not this detailing is not understood and buildings are generally constructed in the same fashion as would clay bricks. They outperform clay bricks in some respects, particularly about frost resistance.

References

[1]   Information on https://theconstructor.org/building/calcium-silicate-bricks-masonry-construction/17256/

[2]   M. K. Sahu, L. Singh, Critical Review on Types of Bricks: Calcium Silicate Bricks, International Journal of Mechanical and Production Engineering, Vol. 5, Issue 11, (2017).

[3]   Information on https://constructionor.com/sand-lime-bricks/

[4]   Information on http://buildingdefectanalysis.co.uk/masonry-defects/an-introduction-to-calcium-silicate-bricks/

CHAPTER 4

Autoclaved Aerated Concrete (AAC) Blocks

1. Introduction

Traditional bricks are the main building materials that are used extensively in the construction and building industry in India. Due to the rapid urbanization and expanding interest for development materials, block furnaces have quickly developed which have legitimately or in a roundabout way caused a progression of ecological and medical issues. At a worldwide level, ecological contamination from block-making activities adds to the wonders of an earth-wide temperature boost and environmental change. The different kinds of squares can be utilized as an option in contrast to the red blocks, to diminish natural contamination and global warming. AAC squares might be one of the answers for block substitution. Like froth concrete, Autoclaved Aerated Concrete (AAC) is one of the confirmed green structure materials, which can be utilized for business, modern and private development. The AAC block is a lightweight material and has high warm protection esteem. It has the basic properties required for use as a structured segment. Because of the lightweight and high solidarity-to-weight proportion of circulated air through solid items, their utilization brings about an obvious economy in the auxiliary individuals, and along these lines spare concrete and steel support. With AAC, the development procedure can be around 20 percent quicker. It weighs just around 50 percent of a standard solid square and has high warm protection and is acoustics-accommodating. It additionally has preferred imperviousness to fire over fly debris and is non-ignitable. It's non-unfavorably susceptible and consequently keeps up the nature of air inside a structure without changing its properties after some time. Utilizing AAC can lessen development costs by about 2.5 percent for structures, for example, schools and medical clinics, and diminish the running expenses of lodging and places of business by 30 to 40 percent after some time. Fig. 1 shows an AAC block, (brand name as Aerocon blocks).

Figure 1. Aerated Autoclaved Concrete (AAC) blocks.

Advances in Bricks and Blocks for Building Construction   Materials Research Forum LLC
Materials Research Foundations **108** (2021)   https://doi.org/10.21741/9781644901519

The Autoclaved aerated concrete (AAC) was created and developed in 1924 by the Swedish architect Dr. Johan Axel Eriksson, along working with Professor Henrik Kreüger at the Royal Institute of Technology [1]. It is an eco-accommodating structural material that originates from modern waste and is produced by using non-poisonous and non-toxic ingredients. With AAC, the development procedure can be around 20 percent quicker. It weighs just around 50 percent of a standard solid square and has high warm protection and is acoustics-accommodating. It additionally has preferable imperviousness to fire over fly debris and is non-burnable. It's non-hypersensitive and henceforth keeps up the nature of air inside a structure without changing its properties after some time. As indicated by one report, AAC now represents over 40% of all development in the United Kingdom and over 60% of development in Germany [2].

2. Raw materials used in manufacturing of AAC blocks

As compared to most other concrete building applications in the construction industry, Aerated Autoclaved Concrete (AAC) is created utilizing nothing bigger than sand. Quartz sand, calcined gypsum, lime (mineral) as well as concrete, and water are utilized as coupling specialists. In certain nations, similar to India and China, fly debris produced from thermal power plants and having 50-65% silica content is used as an aggregate. Many raw materials are used in the manufacturing of AAC blocks. They are summarized in Table 1.

Table 1. Percent proportion of raw materials

S. No.	Material	% of proportion for AAC Blocks with Fly ash / with Sand
1	Flyash / Sand	65-70 / 55-65 %
2	Cement - 53 GRADE OPC	6-15 / 10-20 %
3	Lime	18-25 / 20-30 %
4	Gypsum	3 – 5 / 2 – 3 %
5	Aluminium powder paste (600 kg / m3)	8 %
6	Water	0.6 – 0.65 %
7	Casting system	36-40 ° C / 35-38 ° C

3. Specifications and performance summary of AAC blocks

The product specification and its performance summary of AAC blocks are summarized as below [3]:

3.1   Appearance

The autoclaved aerated concrete (AAC) is light-hued and contains numerous small voids that can be observed when taken a gander at intently. The gas used to 'froth' the solid during the manufacturing process is hydrogen from the chemical reaction from the Aluminum paste with alkaline soluble components in the cement concrete. These air pockets add to the material's protecting properties. In contrast to stonework, there is no immediate way for water to go through the material; be that as it may, it can suck up dampness and a suitable covering is required to forestall water infiltration.

3.2   Size and Density

The Autoclaved aerated concrete (AAC) blocks are made of 625 mm length, 250 mm tallness, and of different thicknesses: 100, 125, 150, 200, 225, 250, 300 mm with a resistance of ±1.5 mm. The thickness of the block is 600 to 650 kg/cum, though the blocks have a thickness of 1750 to 2000 kg/cum. The density of wet blocks is around 800 kg/cum as compared to that of red clay bricks (2400 kg/cum).

3.3   Structural Capability

The compressive quality of Autoclaved aerated concrete (AAC) blocks is excellent. The compressive quality is from 35 to 50 kg/cm2 (according to IS: 2185). Although it is one-fifth the thickness of typical solid it despite everything has a large portion of the bearing quality, and loadbearing structures up to three stories high can be securely raised with AAC blockwork. The AAC is 3-4 times lighter than traditional bricks, therefore, easier and cheaper to transport. Usage reduces the overall dead load of a building, thereby allowing the construction of taller buildings. Entire building structures can be made in AAC from walls to floors and roofing with reinforced lintels, blocks, and floor, wall, and roofing panels available from the manufacturer. AAC floor panels can be used to make non-loadbearing concrete floors that can be installed by carpenters. Lightweight blocks diminish the mass of a structure, along these lines diminishing the effect of tremor on a structure.

3.4   Thermal Mass

The thermal mass performance of Autoclaved aerated concrete (AAC) is dependent on the climate in which it is used. With its mixture of concrete and air pockets, AAC has a moderate overall level of thermal mass performance. Its use for internal walls and flooring can provide significant thermal mass. The temperature moderating thermal mass is most useful in climates with high cooling needs.

3.5   Insulation

The Autoclaved aerated concrete (AAC) has very good thermal insulation qualities relative to other masonry. A 200mm thick AAC wall gives an R-value rating of 1.43 with 5% moisture content by weight. With a 2–3mm texture coating and 10mm plasterboard internal lining it achieves an R rating of 1.75 (a cavity brick wall achieves 0.82). A texture-coated 100mm AAC veneer on a lightweight 70mm or 90mm frame filled with bulk insulation achieves a higher R rating than an otherwise equivalent brick veneer wall (see Insulation; Lightweight framing). Relative to their thickness, The Thermal Conductivity of AAC blocks is 0.16 kW/m°C against 0.70 bricks, thus recurring energy cost is reduced in air conditioning. AAC panels provide less insulation than AAC blockwork, e.g. a 100mm block work AAC wall has a dry state R-value of 0.86 and a 100mm AAC wall panel has a dry state R-value of 0.68.

3.6   Sound insulation

With its closed air pockets, Autoclaved aerated concrete (AAC) can provide very good sound insulation. It has superior sound absorption qualities due to the porous structure of blocks. Combining the AAC wall with an insulated asymmetric air cavity system gives a wall excellent sound insulation property. AAC offers sound attenuation of about 42 dB, blocking out all major sounds and disturbances. The Sound Reduction Index is 45 dB for 200 mm thick block walls (against 50 for 230 mm thick wall). It is ideal for schools, hospitals, hotels, offices, multi-family housing, and other structures that require acoustic insulation.

3.7   Fire and Vermin Resistance

Autoclaved aerated concrete (AAC) is inorganic, incombustible, and does not explode; it is thus well suited for fire-rated applications. Depending on the application and the thickness of the blocks or panels, fire ratings of up to four hours can be achieved. AAC is non-combustible and fire-resistant up to 1600°C. It can withstand up to 6 hours of direct exposure. Due to the structure of blocks, AAC cannot be damaged or infested by termites and other pests. It does not attract rodents or other pests nor can it be damaged by such.

3.8   Durability and moisture resistance

The purposely lightweight nature of Autoclaved aerated concrete (AAC) makes it prone to impact damage. With the surface protected to resist moisture penetration, it is not affected by harsh climatic conditions and does not degrade under normal atmospheric conditions. The level of maintenance required by the material varies with the type of finish applied. The porous nature of AAC can allow moisture to penetrate to depth but the appropriate design (damp proof course layers and appropriate coating systems) prevents this from happening. AAC does not easily degrade structurally when exposed to moisture, but its thermal performance may suffer. Several proprietary finishes (including acrylic polymer-based texture coatings) give durable and water-resistant coatings to AAC blockwork and panels. They need to be treated similarly with acrylic polymer-based coatings before tiling in wet areas such as showers.

3.9   Water absorption

In Autoclaved aerated concrete (AAC) curing takes place at a high temperature and high pressure in saturated steam. During curing, part of the siliceous material (fly ash) reacts chemically with the calcareous ingredients such as lime liberated by the hydration of cement to form a micro-crystalline structure of tobermorite with a much lower specific surface and is characterized by pores formed by the release of H_2 gas during the casting-rising stage of production.

3.10   Toxicity and breathability

The aerated nature of AAC facilitates breathability. There are no toxic substances and no odor in the final product. However, AAC is a concrete product and calls for precautions similar to those for handling and cutting concrete products. It is advisable to wear personal protective equipment such as gloves, eyewear, and respiratory masks during cutting, due to the fine dust produced by concrete products. If low-toxic, vapor-permeable coatings are used on the walls, and care is taken not to trap moisture where it can condense, AAC may be an ideal material for homes for the chemically sensitive.

3.11   Constructability, availability, and cost

Although AAC is relatively easy to work, it is one-fifth the weight of concrete comes in a variety of sizes and is easily carved, cut, and sculpted, it nevertheless requires careful and accurate placement: skilled trades and good supervision are essential. Different sizes of blocks help reduce the number of joints in wall masonry. Lighter blocks make construction easier and faster. It reduces construction time by 20%. They are easy to install. AAC sets and hardens quickly. Blocks can be easily cut, drilled, nailed, milled, and grooved to fit individual requirements. Thick-bed mortar is more forgiving but is uncommon and not the industry preferred option. It also simplifies hydro-sanitary and electrical installations, such as pipes or ducts, which can be installed after the main construction is complete. The construction process with AAC produces little waste as blockwork offcuts can be reused in wall construction.

4.   Autoclaved aerated concrete (AAC) block walling system

The AAC blocks are generally used in wall masonry. The process of laying and construction of a complete wall unit is summarized in the following steps [4].

4.1   Laying of blocks

In laying AAC blocks, the procedure is the same as that for conventional half brickwork except that the block should be slightly wet with a sprinkler before use and not soaked as in the case of bricks.

(a) Before laying the first course the alignment of the wall is marked over the DPC.

(b) The blocks for 1st course should first be laid dry without mortar along a stretched thread between properly located corners of the wall to determine the correct position of the blocks, including those of cross walls joining it, and also to adjust their spacing within the wall length.

(c) When the blocks are set in a proper position, the two corner blocks are removed. A specified mortar bed is spread for the required bed thickness (10mm).

(d) The blocks are laid back in place with true level and plumb.

(e) The thread is then stretched tightly along with the faces of the two corner blocks and the faces of the intermediate blocks are adjusted to coincide with the thread line.

(f) Each intermediate block is removed and re-laid with mortar.

(g) After every three or four blocks, layers are laid, their correct alignment level and verticality are checked.

(h) In the vertical joints, the mortar is applied only on the face of the blocks.

4.2   Electrical and plumbing installations

Electrical and plumbing installations in AAC masonry are placed in routed chases. Care should be taken when laying out chases to ensure that the structural integrity of the AAC elements is maintained. Do not cut reinforcing steel or reduce the structural thickness of the AAC elements except where permitted by the designer. In vertically spanning AAC elements, horizontal routing should be permitted only in areas with low flexural and compressive stresses. In horizontally spanning AAC elements, vertical routing should be minimized.

4.3 Exterior finishes

Unprotected exterior AAC deteriorates when exposed to cycles of freezing and thawing while saturated. To prevent such freeze-thaw deterioration, and to enhance the aesthetics and abrasion resistance to AAC, exterior finishes should be used. They should be compatible with the underlying AAC in terms of thermal expansion and modulus of elasticity and should be vapor permeable. Many different types of exterior finishes are available. Polymer-modified stuccos, paints, or finish systems are the most common exterior finish for AAC. They increase the AAC's water-penetration resistance while allowing the passage of water vapor. Heavy acrylic-based paints containing aggregates are also used to increase abrasion resistance. There is generally no need to level the surface, and horizontal and vertical joints may be chamfered as an architectural feature or may be filled.

4.4 Interior finishes

Interior finishes are used to enhance the aesthetics and durability of AAC. They should be compatible with the underlying AAC in terms of thermal expansion and modulus of elasticity and should be vapor permeable.

(a) Many different types of interior finishes are available. Interior AAC wall panels may have a thin coat of mineral-based plaster to achieve a smooth finished surface. Lightweight interior gypsum-based plaster may provide a thicker coating to level and straighten walls, and to provide a base for decorative interior paints or wall finishes. Interior plasters have bonding agents to enhance their adhesion and flexibility and are commonly installed by either spraying or troweling.

(b) For commercial applications requiring high durability and low maintenance, acrylic-based coatings are often used. Some contain aggregates to enhance abrasion resistance.

(c) When ceramic wall tile is to be applied over AAC, surface preparation is normally necessary only when the AAC surface requires leveling. In such cases, a Portland cement or gypsum-based parge coat is applied to the AAC surface before setting the ceramic tile. The ceramic tile should then be adhered to the parged wall using either a cement-based thin-set mortar or an organic adhesive. In moist areas such as showers, only a Portland cement-based parge coat should be used, and the ceramic tile should be set with cement-based, thin-set mortar only.

4.5   Service lines

For concealed or piping, a block wall can be chased using a hand or electric router. Depths of vertical chases should be limited to one-third of the wall thickness and horizontal chases to one-sixth of the wall thickness. Holes in a block wall can be made with a standard hand or electric drill. The chases shall be refilled with a leaner mortar and chicken mesh shall be applied to that area and cured (Figure 2).

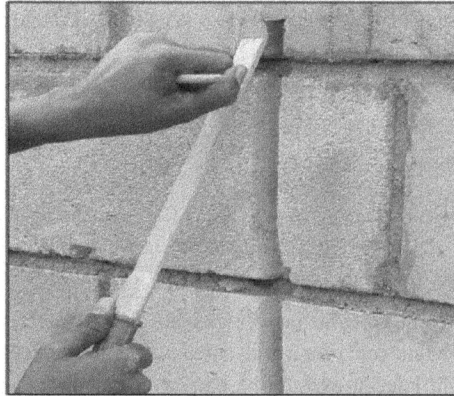

Figure 2. The chasing of the wall for service lines

4.6   Lintels

Precast or cast-in-situ concrete lintels can be used in block masonry, over all openings. Lintels shall always rest on a full block with a minimum bearing as under. Below the openings, the RCC band should be provided with reinforcement to avoid diagonal tension cracks. The bond beam to be extended up to 300 mm from window corners on both sides. Table 2 shows the minimum bearing on each side for the different opening sizes.

Table 2. Minimum bearing on each side for different opening sizes.

Opening Size	Up to 900 mm	900 to 2000 mm	2000 to 3000 mm	Above 3000 mm
Minimum bearing (each side)	150mm	200mm	300mm	To design

4.7 Plastering

Followings are the points that should be considered while plastering the AAC walls:

(a) Do not soak the wall before plastering. The wall shall be moistened evenly before applying the plaster. A fog spray is recommended for this purpose.

(b) For external plastering has to be carried out in two coats, apply SBR coating with sand on the block surface will enhance the bonding and minimize the thickness of plastering.

(c) It is recommended to use cement mortar 1:6 for internal & external plastering works and preferably use PPC cement for masonry and plastering works to minimize shrinkage cracks.

(d) Plastering thickness can be minimized to 10 mm and 15 mm for internal and external walls.

4.8 Precautions while laying the AAC masonry blocks

The cracks occurring in block masonry and plastering are of any structural problem involving stability and safety to the structure. But it is advised to minimize the same to have a good appearance and maintenance-free.

(a) Do not store the blocks on an unleveled surface

(b) Do not use wet blocks for masonry construction

(c) Do not make the holes on block masonry for scaffolding supports

(d) Do not soak the blocks before use

(e) Do not hammer the block masonry for service lines, chases, etc.

(f) Do not completely wet the block masonry before plastering works

(g) Do not chase the blocks back to back for lesser thickness blocks.

5. Advantages of using AAC blocks

Autoclaved aerated concrete (AAC) has many advantages as compared to other cement concrete materials. The basic advantage is that it is cost-efficient and eco-friendly with having a low environmental impact. In the manufacturing of AAC blocks, the topsoil of the earth's surface is not used, therefore it emits very low carbon dioxide as compared to red clay bricks. Since AAC blocks are an industrial product manufactured with machines, the quality of the end products is very good, uniform, and consistent. The blocks are even and well finished from all sides, therefore the thickness of wall plaster is reduced. Due to its lightweight, there will also be a reduction of dead weight in the structural system, which will lead to the saving of steel and concrete. Also, this leads to the construction of more stories or taller buildings. The economy is achieved in multistoried, especially buildings by using AAC blocks as compared to red clay bricks due to a significant reduction in the dead weight of the structure; and hence it reduces the cost of the RCC framed structure.

The AAC blocks have a lot of pores and voids; hence it provides better sound absorption and insulation as compared to red clay bricks or concrete blocks. The AAC blocks have a low thermal conductivity which is approximately 0.24 kW-M/°C, which results in the saving of electricity costs by 30%, which is required for heating and cooling of the house. The manufacturing process of AAC blocks is non-polluting. The by-product from the manufacturing industry is only steam. All the ingredients to manufacture the AAC blocks are non-toxic and safe. The AAC blocks are fire-resistant and non-combustible. They can tolerate up to 6 hours of direct exposure to fire. The AAC blocks have air voids and hence have better fire-resisting properties compared to red clay bricks. The melting point of the AAC blocks is over 1600 °C more than twice the typical temperature in building fire 650 °C. The AAC blocks are easy to work with i.e they can be easily cut nailed and drilled and can be fitted to the individual requirements which gives more design flexibility. It cannot be rotten easily since it is pest and termite-proof. The AAC walling system has simplified hydro-sanitary and electrical installations, such as pipes or ducts, which can be installed after the main construction is completed. It also has low maintenance and it reduces the operating and maintenance cost by 30% to 40%. The chemical mortars can be used for joining the AAC blocks in masonry work; hence this reduces the material consumption for cement mortar and also there is no need for curing. The AAC blocks are suitable both for non-load bearing strictures and also for reinforced cement concrete structures in partition walls.

6.   Disadvantages of AAC blocks

There are many advantages of using AAC blocks, which has made this material a very suitable and sustainable material in the building and construction sector. But there are also a few disadvantages of using AAC blocks. It has been observed that the Aircrete cracks after installation in the rainy season, which can be avoided by reducing the strength of the mortar and ensuring the blocks, are dry during and after installation. The AAC blocks should be handled carefully then as compared to red clay bricks to avoid breakages since it is somewhat brittle. Due to its brittle nature, it requires long thin screws when fitting cabinets and wall hangings. The initial cost of the manufacturing industry is a bit high. There are not many factories, which are producing AAC blocks; therefore it is not very easily available.

7.   Comparative analysis of AAC blocks, clay bricks, and CLC blocks

The comparative analysis of Autoclaved Aerated Concrete (AAC) blocks, Red clay bricks, and Cellular Light Weight Concrete (CLC) blocks are summarized in Table 3:

Table 3: Comparison of AAC Block, Clay Bricks, and CLC Blocks [5].

S. No.	Parameter	AAC Blocks	Clay Red Bricks	CLC Blocks
1	Raw Materials	Cement, fly ash, water, and Air entraining agents	Locally available clay	Cement, lime, especially ground sand, foam
2	Size	400-600mm X 200mm X 150mm – 300mm	225mm X 75mm X 100/150mm	400-600 x 200 x 100/150/200 mm
3	Variation Size	1.5 mm (+/-)	5 mm (+/-)	5 mm (+/-)
4	Compressive Strength (As per IS codes)	3-4 N/mm2	3.5 N/mm2	2 -2.5 kg/cm2
5	Dry Density (As per IS codes)	550-650 kg/m3 Its one-third of the weight of clay brick which makes it easy to lift and transport	1800 kg/m3	800 kg/m3
6	Cost-Benefit	For high-rise buildings, there is the reduction of Deadweight which leads to saving in Concrete and steel.	As easily available in the local market hence it is beneficial for low rise structure.	For high-rise buildings, there will be a reduction of Deadweight which leads to saving in Concrete and steel quantities.
7	Fire Resistance (8″ Wall)	Up to 4 Hours	Around 2 Hours	Around 4 Hours
8	Quality of End Product	Factory-made product. So the quality of the end product is consistent and good	Locally made products. Quality depends on various parameters like raw materials quality, the manufacturing process, etc.,	The quality of the end product depends on the foam used and the degree of quality control

9	Sound Insulation	Better Sound absorption /insulation as compared to bricks	Normal	Better Sound absorption /insulation as compared to bricks
10	Energy Saving	Low thermal conductivity (0.24 Kw-M/C) helps in saving electricity costs 30% for heating and cooling.	High thermal Conductivity (0.81 Kw-M/C). So no significant cost savings	Low thermal conductivity (0.32 Kw-M/C) helps in saving electricity costs 30% for heating and cooling.
11	Environmental Friendliness	In AAC Block there is no topsoil consumption and it emits very low CO_2 as compared to Red clay bricks while manufacturing	One sq ft of carpet area with clay brick walling will consume 25.5 kg of topsoil (approx.). It damages the environment	In CLC Block there is no topsoil consumption and it emits very low CO_2 as compared to Red clay bricks while manufacturing.
12	Internal and External Plaster	As these bricks have dimensional accuracy, the internal and external plaster thickness is reduced	Requires thick plaster surface as there are variations in the dimensions	As these bricks have dimensional accuracy, the internal and external plaster thickness is reduced
13	Cost of Construction (year 2018)	1 Cum costs – Rs. 4200/-	1 Cum costs – Rs. 2440/-	1 Cum costs – Rs. 4000/-
14	Joining Process	Chemical mortars can be used for joining the brick. This reduces the material consumption for cement and avoids the curing process	The traditional mortar needs to be used and the brickwork should be cured at least for 7 days before plastering	Chemical mortar can be used for joining the brick. This reduces the material consumption of cement & avoids the curing process.
15	Availability	The factory setup cost is high. Not many factories, so availability is a concern.	Available locally in all cities and villages.	Factory setup cost is low as compared to AAC. It also takes a long time to produce if steam curing is not used. Timely availability is a concern.

16	Thermal Insulator	AAC Blocks are good thermal insulator, monthly expenses it will save cost for an entire lifetime	It has low thermal insulation as compared to AAC and CLC Block	CLC Blocks are good thermal insulators. monthly expenses it will save cost for an entire lifetime
17	Tax Contribution	Contributes to Government taxes in form of Central, Excise, and VAT	No Tax Contribution	Contributes to Government taxes in form of Central Excise and VAT
18	Cylindrical Structures	For Cylindrical structure, these blocks are not much useful	Cylindrical manholes need small size of bricks so that the curvature can be formed hence Red clay bricks are useful	For Cylindrical structure, these blocks are not much useful
19	Water Absorption	Absorb 12- 15% by the total volume of AAC blocks	Absorb 17 -20% by the total volume of red clay brick	Absorb 12-15% of water by the total volume of Block
20	Range of Application	They are suitable for non-load-bearing or RCC structure in the partition wall	They are useful in both load-bearing and non-load bearing structure	They are suitable for non-load-bearing or RCC structure in the partition wall

8.   Cost comparative analysis of AAC blocks, Clay Bricks, and CLC blocks

The cost comparative analysis of brickwork in masonry and plaster for AAC blocks, Red clay bricks, and Cellular Light Weight Concrete (CLC) blocks are summarized in Table 3 and Table 4 respectively [6].

Table 3: Cost comparative analysis for AAC blocks and clay brick masonry for 1M³ [1:4]

S. No.	Parameters	Clay Red Bricks	AAC Blocks
1	Quantity Analysis	200mmx 100mm x 100mm	600mm x 200mm x 200mm
2	No. of bricks / blocks	500	37
3	Mortar Quantity	0.2766 M³	0.1344 M³
4	No. of begs of cement	1.65	1.0
5	Quantity of Sand	0.221M³	0.1075 M³
6	Quantity of Water	31 Litres	16 Litres
7	Rate Analysis	5252.00 Rs./ m² (As per MP PWD SOR building work 2014 clause no.6.3)	5052.00 Rs. / m² (As per MP PWD SOR building work 2014 clause no.6.27)

Table 4: Cost comparison for plasterwork for AAC blocks and clay bricks for 1 M³ [1:4]

S. No.	Parameters	Clay Red Bricks	AAC Blocks
1	The volume of mortar for plaster	1.8 M³	1.0 M³
2	The volume of mortar by 25% for wastage and frog filling	2.25 M³	1.25 M³
3	Quantity of cement	0.45 M³	0.25 M³
4	No. of begs of cement	13.5	7.5
5	Quantity of Sand	1.8 M³	1.0 M³
6	Quantity of Water	236.25 Liters	131.25 Liters
7	Rate Analysis	171.00 Rs./ m² (As per MP PWD SOR building work 2014 clause no.13.6)	91.10.00 Rs./ m² (As per MESSOR building work 2010 item no.14001)

9.   Environmental benefits of AAC blocks

The Autoclaved Aerated Concrete (AAC) is an eco-friendly material that has many environmental benefits. The weight of the AAC block is around one-fourth to one-fifth that of concrete based on volume. The manufacturing of AAC blocks has the same greenhouse gas environmental impact and has the same embodied energy as that of concrete blocks. The AAC blocks or panels have lower embodied energy per square meter than a concrete alternative building material. The AAC block and panels have more insulation value and thus it has low energy usage for heating and cooling loads requirement. The total energy used in manufacturing the ACC blocks is around 50% less than that of manufacturing other prefabricated building components and products. As compared to regular cement concrete building products, AAC reduces around one-third of the environmental waste. The Autoclaved Aerated Concrete (AAC) blocks and panels have proven to be more durable, provide thermal insulation and structural requirements, and also have major economic and environmental benefits as compared to other traditional building components and products. Thus Autoclaved aerated concrete can be said to a suitable and potential eco-friendly building material, which is beneficial for the environment, which fulfills the requirement for the construction of sustainable architecture and construction [7].

Conclusion

The Autoclaved Aerated Concrete (AAC) is a novel and one of the most suitable and sustainable building materials in the present building construction industry. AAC blocks are a result of productive use of recycled industrial waste i.e. fly ash, hence this material can be classified as a sustainable building material. The production price of AAC blocks at the manufacturing unit is Rs. 3200/- to Rs. 3600/ per cubic meter as per the rates in India in the year 2019. The inherent properties of AAC blocks result in fast and efficient construction techniques. Hence the Autoclaved Aerated Concrete (AAC) has become an efficient building construction material that is being used in a wide range of residential, commercial, and industrial buildings and it has been used in the Gulf countries for the last 40 years and in Europe for since 70 years, and in Australia and South America for the past 20 years. According to a report, the AAC blocks are used in more than 60% of construction in Germany, and England approximately 40% of all construction industry [8]. Since the AAC blocks use readily available raw materials in the manufacturing process, have excellent durability, are energy efficient, are cost-effective, and also can be recycled, therefore Autoclaved Aerated Concrete (AAC) can be said to be a green and sustainable building material.

References

[1] Five green building blocks, information on http://www.thehindu.com/features/homes-and-gardens/5-green-building-blocks/article4813910.ece

[2] S. Schnitzler, Autoclaved Aerated Concrete as a Green Building Material, UC Davis Extension, Switzerland, 2016.

[3] M. Arif Kamal, Autoclaved Aerated Concrete (AAC): A Sustainable Building Material information on http://www.masterbuilder.co.in/autoclaved-aerated-concrete-aaca-sustainable-building-material/

[4] New Building Materials and Technologies, Vol. IV, Compendium of New Building Technologies, Indian Building Congress, New Delhi, India, 2019.

[5] Comparison of AAC Blocks vs CLC Blocks vs Red Clay Bricks, information on https://happho.com/comparision-aac-blocks-vs-clc-blocks-vs-red-clay-bricks/

[6] U. Jain, M. Jain, S. Mandaokar, Comparative Study of AAC Blocks and Clay Brick and Costing, International Journal of Research in Engineering, Science and Management Vol. 1, Issue 9, Sept. 2018, India.

[7] Global Autoclaved Aerated Concrete Market Outlook: Trend and Opportunity Analysis, Competitive Insights, Actionable Segmentation and Forecast 2023, Research Report, Energias Market Research, 2019, Buffalo, USA.

[8] Autoclaved Aerated Concrete (AAC) A Sustainable Building Material, to Witness a CAGR of 7.9% during 2017 – 2023, information on

https://www.globenewswire.com/news-release/2018/02/23/1386502/0/en/Autoclaved-Aerated-Concrete-AAC-A-Sustainable-Building-Material-to-Witness-a-CAGR-of-8-0-during-2018-2024-Energias-Market-Research-Pvt-Ltd.html

CHAPTER 5

Compressed Earth Blocks

1.   Introduction

Soil is the most abundant construction material known to man. Mud block wall construction with a different type of stabilization was predominant in earlier days. A compressed earth block (CEB), also known as a pressed earth block or a compressed soil block, is a building material made primarily from damp soil compressed at high pressure to form blocks. Compressed earth blocks use a mechanical press to form blocks out of an appropriate mix of fairly dry inorganic subsoil, non-expansive clay, and aggregate. If the blocks are stabilized with a chemical binder such as Portland cement they are called compressed stabilized earth block (CSEB) or stabilized earth block (SEB). Typically, around 3,000 psi (21 MPa) is applied in compression, and the original soil volume is reduced by about half.

During the last century, this type of block was neglected due to the advent of cement and the increasing production of burnt clay bricks. Due to the excessive energy consumed in their production now there is an upsurge of interest in stabilized soil blocks. The top layer of soil, which is called the "topsoil", usually contains a considerable proportion of organic matter as a result of the accumulation of mineral and vegetable remains. Beneath the topsoil lies subsoil, containing little or no organic matter and is suitable for use as a building material. Creating CEBs differs from rammed earth in that the latter uses a larger formwork into which earth is poured and manually tamped down, creating larger forms such as a whole wall or more at one time rather than building blocks. CEBs differ from mud bricks in that the latter are not compressed and solidify through chemical changes that take place as they air dry. The compression strength of properly made CEB usually exceeds that of typical mud brick. Building standards have been developed for CEB. CEBs are assembled onto walls using standard bricklaying and masonry techniques. The mortar may be a simple slurry made of the same soil/clay mix without aggregate, spread or brushed very thinly between the blocks for bonding, or cement mortar may also be used for high strength, or when construction during freeze-thaw cycles causes stability issues [1].

2.   Basic components

Soil is composed of several basic elements, in varying proportions: very coarse sand, coarse, medium, and fine sand, silt, and clay. The usual method of determining the proportions is to classify the soil by particle size. Conventionally, soils are classified based on their granulometric analysis and according to the name of the predominant basic component is sandy, silty, or clayey.

To facilitate classification, however, the basic soil components have been grouped according to particle size, as follows:-

- Coarse particles, or sand, from 4.76 mm to 0.074 mm in size
- Fine particles (silt and clay), less than 0.074 mm in size (ASTM No. 200 sieves)

2.1   Coarse particles

Sand: This can be regarded as the structural and inert element in soil. It is stable when moist, but unstable when dry. Sand has high internal friction. It has no cohesion and no plasticity. It does not contract when dried. It is pervious and is compressed almost instantaneously when pressure is applied to its surface.

2.2   Fine particles

Silt: The size of the silt particles cannot be seen with the naked eye, but they feel gritty between the teeth. Silt has little cohesion of internal friction. It may change in volume when worked. It is difficult to compact.

Clay: This is the type of soil, with important properties, both physically and chemically. Because of its high plasticity, clay readily takes the desired shape. It is a material smooth to the touch and sticky when wet. Dry clay absorbs a considerable proportion of water with a notable increase in volume; when it dries, it returns to its original volume, but the shrinkage of the mass results in the formation of cracks. It is virtually impervious and is compressed very slowly when pressure is applied to its surface.

To improve mud structures a method adopted is known as soil stabilization. Soil stabilization is the name given to certain processes to which natural soils are subjected to make them more suitable for use as a construction material. The addition of stabilizing agents not only enhances the best qualities of soils but imparts to the other properties which they alone do not possess. The cost of soil stabilization is very low when compared to its gains in terms of value addition to its benefits using as a building material.

3.   Addition of a binding agent

This method consists of mixing soil with some agent so that its particles bind together and remain firmly united and unaffected by moisture variations, thus producing a strong and very durable material. The binding agent generally used is Portland cement. It is a compound of soil, cement, and water. The high quality of the resulting construction material depends on the use of correct proportions of soil, cement, and water. Table 1, below mentions the percentage of cement required.

Table 1.   The percentage of cement required for different types of soil.

Type of soil	Normal percentage of cement
Sandy	4.75 – 9.10
Silty	8.35 – 12.5
Clayey	– 15.4 (not recommended for use)

Each type of soil requires a specific degree of moistness if it is to be compacted correctly. Broadly speaking, the total amount of water varies between 8 percent and 16 percent by volume.

4.   Block making machines

Currently, large varieties of machines are commercially marketed throughout the world for the production of stabilized soil blocks. For small production at present in India, three types of manually operated machines are available for making soil cement blocks. They are ASTRAM, ITGE VOTH, and BALRAM. Generally, these machines consist of a mold to receive the soil, a toggle lever mechanism to compact the soil, and a frame to support the mold and the lever. The mold is provided with a stiffened plate lid which can be locked down after closing. The mold is interchangeable and generally, molds of two sizes are used. The two molds can produce blocks of sizes 300 x 145 x 100mm and 23 x 190 x 100mm respectively. A scoop is provided to fill the soil into the mold which is to be used in measuring out the right amount of soil [2].

Either hand-operated or motor-driven molders may be used for making blocks by this method. For large production hydraulic operated machine is used for compacting soil cement into blocks of the desired size.

Figure 1, 2 and 3 shows few machines for mechanized production

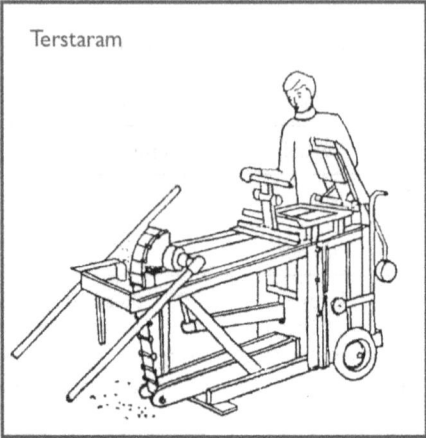

Figure 1.   A manually operated block making machine.

Figure 2.   A semi-mechanized portable machine.

Figure 3.   An automatic block making machine.

5.   Manufacturing process

The manufacture of stabilized soil blocks will be as per IS code 1725. Large numbers of buildings using stabilized soil blocks have come up in many parts of the world including India. This is mainly because stabilized soil block is energy efficient and low carbon emission. The soil blocks are alternative to burnt clay brick and concrete-based blocks.

The process of manufacture of soil-cement blocks involves the following five steps:

- Analysis of the soil
- Sifting of the soil
- Preparation of the mix
- Compaction of the blocks
- Curing of the blocks

5.1   Analysis of the soil

Soils are made up of three components: sand, silt, and clay. These components are defined based on particle size, and is the coarsest of the three, and clay the finest. A well-graded soil gives good results. In general, soil consisting of 33% fine, 33% silt, and 33% sand is good for blocks. The mixing of 5 to 10% cement by volume offers comparatively good bricks/blocks.

5.2 Sifting of the soil

Soil should be dried and sieved (to remove large lumps, stones, leaves, and other impurities) before it can be used properly mixed with cement and compressed into blocks. The sturdy frames with metallic meshes can be used for the sifting soil.

5.3 Preparation of the mix

Once the soil has been dried and sifted, one can begin to prepare the mix from which blocks will be pressed. The amount of Portland cement to be used will depend on the composition of the soil. Sandy soils require 5 to 9% cement by volume. Silty soils need 8 to 12%, and clayey soils require 12 to 15% cement as a stabiliser. More than 15% by volume is not recommended. Mix thoroughly all the ingredients: cement, soil, and special additions such as sand or clay that may be needed. After dry mixing of all the ingredients, water is added a little at a time until the damp soil cement reaches the right consistency. The mixing should be continued till the mixture becomes uniform in color and appearance. It is suggested that blocks should be cast within half an hour of mixing otherwise strength and durability of the blocks decreases if the compaction of soil-cement is delayed further.

5.4 Compaction of the blocks

The prepared mix is placed into the mold of the machine and pressure is applied. After compaction, the block formed is ejected from the mold and stacked. As the blocks are fragile when newly formed, care is needed while handling and stacking.

5.5 Curing of the blocks

Place the blocks as soon as possible on a flat, non-absorbent surface in a shady environment with protection from sun and rain to cure. Set each block on edges and space the blocks far enough apart so that they do not touch each other. The blocks must be allowed to dry slowly and without violent changes in temperature. Loss of moisture must therefore be strictly controlled during the first twenty-four hours after they are made, to prevent them from drying out completely all at once, since this may affect the quality of the material. After 24 hours of de-molding, blocks must be thoroughly sprinkled three times a day with the fine water spray for 15 days. The slower the block dries, the stronger it will be in strength (Figure 4).

Figure 4.   *Stacking curing and protection of earth stabilized blocks*

6. Size of blocks

The modular sizes of stabilized soil blocks are given in Table 2. The non-modular sizes are also feasible with appropriate molds.

Table 2.   *The modular sizes of stabilized soil blocks.*

Length (mm)	Width (mm)	Height (mm)
290	90	90
290	140	90
240	240	90
190	90	90
190	90	40

7. Walling with soil cement blocks

7.1 Bonding of the blocks

How blocks are arranged in a wall is called bonding. An essential requirement of correct bonding is that all the joints should be carefully arranged to ensure the proper transmission of vertical loads. In other words alternate courses are so laid that the vertical joints or seams in one course do not coincide with those in the course below.

7.2   Mortar requirement

The strength of a wall depends on the combined strength of the blocks and the mortar. The blocks of the strength of 30 to 60 kg/cm^2 can be produced. If blocks and mortar are of equal strength in a wall subjected to vertical loads, both will bear the pressure equally; but if there is any weakness in the mortar, the blocks will be subjected to shearing stress resulting inevitably in cracks and fissures.

It is a mistake to use high-strength soil-cement blocks with weak mortars or low-strength blocks with good-quality mortars, since in the latter case, conversely to what occurs in the former, the failure will be in the blocks. Absorption of water by the blocks from the fresh mortar is a factor to be reckoned with in soil-cement wall construction. To ensure a sufficient amount of water for the proper setting of the mortar, it is advisable to moisten the surface of the blocks in contact with the mortar.

7.3   Fixing of frames

Concrete –in-situ blocks are used to hold the door / window frame holdfasts, like in a brick masonry wall with burnt bricks.

Figure 5. Institutional building made from compressed earth blocks in New Delhi, India.

8.   Earthquake resistance of blocks

Soil cement is, in general, capable of withstanding certain shocks or stresses caused by seismic disturbances. In regions subject to very intense seismic disturbances, tensile stresses which cannot be absorbed by the soil-cement material may occur; it is, therefore, desirable to provide vertical reinforcement, to absorb stresses of this kind. Adequate vertical reinforcement is provided at wall junctions, as per IS:4326. In seismic regions, it is important to provide at the top of the wall a continuous beam/band which provides rigidity and contributes to the stability of the house. A lintel band is also laid as per IS 4326.

9.   Energy savings

Energy consumed in manufacturing 1000 bricks and that in manufacturing cement used in an equivalent volume of cement stabilized soil blocks using 8% cement have been worked out. Energy used in manufacturing 1000 bricks is 6.90×10^5 KCal while as that used in manufacturing approx. 320 Kgs of cement needed to stabilize the equivalent volume of soil blocks which works out to 2.71×10^5 Kcal. Thus there is a gross saving of 60.72% in energy consumed.

10.   Advantages of compressed earth blocks

The Buildings constructed of compressed earth blocks exhibit numerous benefits.

- Compressed earth is a versatile building material and suitable soil is readily available locally at many places.

- Earth is a fire-resistant material with a 4-hour rating.

- Earth can be used to produce load-bearing structurally sound walls.

- Earth construction – having a large thermal mass – produces a consistent internal temperature. Specifically, it evens out diurnal variations and traps internally generated heat.

- Built environments using earth are quiet, having low sound transmission. Additionally, they have good air quality.

- Earthen structures are durable and resistant to weathering. When treated with water-resistant coating, earthen walls are water-resistant. They resist habitation and damage from insects, such as ants and termites.

- Additional structural elements can be incorporated into earthen walls to provide earthquake protection.

- Economic benefits include low cost of maintenance, low cost of heating and cooling the interior. The construction cost using existing compressed earth block technology is also less. Stabilization can be achieved using several methods: compaction/densification, granular stabilization, or chemical stabilization.

Advances in Bricks and Blocks for Building Construction   Materials Research Forum LLC
Materials Research Foundations **108** (2021)   https://doi.org/10.21741/9781644901519

References

[1]   Information on https://en.wikipedia.org/wiki/Compressed_earth_block

[2]   P. K. Adlakha, New Building Materials, and Technologies, Vol. IV, Indian
Building Congress, New Delhi, India, 2019.

CHAPTER 6

Stabilized Mud Blocks

1. Introduction

Earth block is a construction material made primarily from soil. Types of earth blocks include compressed earth block (CEB), compressed stabilized earth block (CSEB), and stabilized earth block (SEB). Stabilization is a technique of improving the properties of mud in such a way that it will possess adequate wet strength, durability, and dimensional stability (retains its shape and size both in dry and moist conditions) without burning. Compacting a mixture of sandy soil with 7% cement and small quantities of lime at optimum moisture in a press results in stabilized mud blocks (SMB). The overall energy consumption in SMB is quite small (about one-third) in comparison with the burnt bricks. Compaction of the soil-stabilizer mix can be done by using a manually operated or a mechanized press. The press should be capable of generating sufficient force to produce a dense block. It is advantageous to use a manually operated press to eliminate additional energy needs in terms of electricity or diesel. The blocks have to be cured for three weeks by keeping the block surfaces moist. Additives like fly ash, quarry dust from crushers, granite fines, or stone dust from stone cutting and polishing industries and various other mine or factory wastes can be used effectively along with the natural soil [1].

Stabilized mud block (SMB) or pressed earth block is a building material made primarily from damp soil compressed at high pressure to form blocks. If the blocks are stabilized with a chemical binder such as Portland cement they are called compressed stabilized earth block (CSEB) or stabilized earth block (SEB). Creating SMBs differs from rammed earth in that the latter uses a larger formwork into which earth is poured and manually tamped down, creating larger forms such as a whole wall or more at one time rather than building blocks and adobe which is not compressed. Stabilized mud block uses a mechanical press to form a block out of an appropriate mix of fairly dry inorganic subsoil, non-expansive clay, aggregate, and sometimes a small amount of cement (Figure 1).

Typically, around 3000 psi is applied in compression, and the original soil volume is reduced by about half. The compression strength of properly made SMB can meet or exceed that of typical cement or adobe brick. Building standards have been developed for SMB.

Figure 1. Stabilized mud blocks

2.   Soil Stabilization methods with different materials

The following are the various soil stabilization methods and materials:

2.1   Soil stabilization with cement

The soil stabilized with cement is known as soil cement. The cementing action is believed to be the result of chemical reactions of cement with siliceous soil during hydration reaction. The important factors affecting the soil-cement are the nature of soil content, conditions of mixing, compaction, curing and admixtures used.

The appropriate amounts of cement needed for different types of soils may be as follows:

Gravels – 5 to 10%

Sands – 7 to 12%

Silts – 12 to 15%, and

Clays – 12 – 20%

The quantity of cement for compressive strength of 25 to 30 kg/cm^2 should normally be sufficient for tropical climate for soil stabilization.

If the layer of soil having a surface area of A (m^2), thickness H (cm), and dry density r_d (tonnes/m^3), has to be stabilized with p percentage of cement by weight based on dry soil, cement mixture will be

$$\frac{100 \times p}{100 + p}$$

and, the amount of cement required for soil stabilization is given by

Amount of cement required, in tonnes $= \left(\dfrac{AHr_d}{100}\right) \times \left(\dfrac{p}{100+p}\right)$

Lime, calcium chloride, sodium carbonate, sodium sulphate, and fly ash are some of the additives commonly used with cement for cement stabilization of soil.

2.2   Soil stabilization using lime

Slaked lime is very effective in treating heavy plastic clayey soils. Lime may be used alone or in combination with cement, bitumen, or fly ash. Sandy soils can also be stabilized with these combinations. Lime has been mainly used for stabilizing the road bases and the subgrade.

Lime changes the nature of the adsorbed layer and provides pozzolanic action. The plasticity index of highly plastic soils is reduced by the addition of lime with soil. There is an increase in the optimum water content and a decrease in the maximum compacted density and the strength and durability of soil increases.

Normally 2 to 8% of lime may be required for coarse-grained soils and 5 to 8% of lime may be required for plastic soils. The amount of fly ash as an admixture may vary from 8 to 20% of the weight of the soil.

2.3   Soil stabilization with bitumen

Asphalts and tars are bituminous materials that are used for the stabilization of soil, generally for pavement construction. Bituminous materials when added to soil, it imparts both cohesion and reduced water absorption. Depending upon the above actions and the nature of soils, bitumen stabilization is classified into the following four types:

- Sand bitumen stabilization
- Soil Bitumen stabilization
- Waterproofed mechanical stabilization, and
- Oiled earth.

2.4   Chemical stabilization of soil

Calcium chloride is hygroscopic and deliquescent is used as a water-retentive additive in mechanically stabilized soil bases and surfacing. The vapor pressure gets lowered, surface tension increases, and the rate of evaporation decreases. The freezing point of pure water gets lowered and it results in the prevention or reduction of frost heave.

The depressing the electric double layer, the salt reduces the water pick up and thus the loss of strength of fine-grained soils. Calcium chloride acts as a soil flocculent and facilitates compaction.

Frequent application of calcium chloride may be necessary to make up for the loss of chemicals by leaching action. For the salt to be effective, the relative humidity of the atmosphere should be above 30%.

Sodium chloride is the other chemical that can be used for this purpose with a stabilizing action similar to that of calcium chloride.

Sodium silicate is yet another chemical used for this purpose in combination with other chemicals such as calcium chloride, polymers, chrome lignin, alkyl chlorosilanes, siliconites, amines, and quarternary ammonium salts, sodium hexametaphosphate, phosphoric acid combined with a wetting agent [2].

3.   Identification of soil

Very few laboratories can identify soils for building purposes. But soil identification can be performed by anybody with sensitivity analyses. The main points to examine are:

- Grain size distribution, to know the quantity of each grain size
- Plasticity characteristics, to know the quality and properties of the binders (clays and silts)
- Compressibility, to know the optimum moisture content, which will require the minimum of compaction energy for the maximum density.
- Cohesion, to know how the binders bind the inert grains.
- Humus content, to know if they are organic materials which might disturb the mix.

4.   Soil stabilization

Many stabilizers can be used. Cement and lime are the most common ones. Others, like chemicals, resins, or natural products can be used as well. The selection of a stabilizer will depend upon the soil quality and the project requirements. Cement will be preferable for sandy soils and to achieve quickly a higher strength. Lime will be rather used for very clayey soil but will take a longer time to harden and to give strong blocks.

4.1 Soil suitability and stabilization

Not every soil is suitable for earth construction and CSEB in particular. But with some knowledge and experience, many soils can be used for producing CSEB. Topsoil and organic soils must not be used. Identifying the properties of soil is essential to perform, in the end, good quality products. Some simple sensitive analyses can be performed after a short training. Soil is an earth concrete and a good soil for CSEB is more sandy than clayey. Figure 2 shows the proportions of compressed stabilized earth blocks.

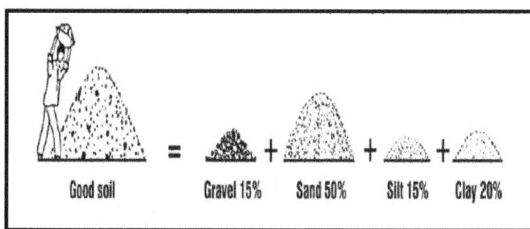

Figure 2. The Proportions of compressed stabilized earth blocks

According to the percentage of these 4 components, a soil with more gravel will be called gravely, another one with more, sand, sandy, others silty or clayey, etc. The field tests aim to identify in which of these four categories the soil is. From the simple classification, it will be easy to know what to do with this soil.

5. Raw materials

Soil characteristics and climatic conditions of an area must be evaluated before manufacturing soil building blocks. A dry climate, for example, needs different soil blocks from those used in temperate, rainy, or tropical areas. All soils are not suitable for every building need. The basic material, however, required to manufacture compressed stabilized earth building blocks is soil containing a minimum quantity of silt and clay to facilitate cohesion. Soils are variable and complex materials, whose properties can be modified to improve performance in building construction by the addition of various stabilizers. All soils consist of disintegrated rock, decomposed organic matter, and soluble mineral salts. Soil types are graded according to particle size using a system of classification widely used in civil engineering. This classification system, based on soil fractions shows that there are 4 principal soil fractions - gravel, sand, silt, and clay.

For soil stabilization, the clay fraction is most important because of its ability to provide cohesion within the soil. The manufacture of good quality, durable compressed stabilized earth blocks requires the use of soil containing fine gravel and sand for the body of the block, together with silt and clay to bind the sand particles together. An appropriate type of stabilizer must be added to decrease the linear expansion that takes place when water is added to the soil sample.

6.   Preparations of raw materials

6.1   The requirements for preparation

The basic materials required for the production of compressed stabilized earth building blocks are soil, stabilizer, and water. The stabilizer, whether lime or cement, or some other material, is usually available in powder or liquid form, ready for use. The soil may be wet or dry when it is first obtained, and will probably not be homogeneous. To have uniform soil, it is often necessary to crush it so that it can pass through a 5 to 6mm mesh sieve. The different soil types may also need to be used together to obtain good quality products. For instance, heavy clay may be improved by the addition of sandy soil. It is not only important to measure the optimum proportion of ingredients, but also to mix them thoroughly. Mixing brings the stabilizer and soil into direct contact, thus improving the physical interactions as well as the chemical reaction and cementing action. It also reduces the risk of uneven production of low-quality blocks. Various types and sizes of mixing equipment are available on the market.

6.2   Breaking up of soil

In most developing countries the soil is usually dry when dug out of the borrow pit or it will dry soon after digging. Simple hand tools are available as well as a range of more complicated machinery that can be used to reduce the soil grains to an appropriate size. To obtain a uniform mix of the mineral components, water and stabilizer lump more than 200mm in diameter after excavation must be broken up. Grains with a homogeneous structure, such as gravel and stones, must be left intact, and those having a composite structure (clay binder) broken up so that at least 50 percent of the grains are less than 5mm in diameter. The soil must be dry as wet soil can only be handled by certain mechanized systems.

6.3   Grinding followed by screening

The material is pressed between two surfaces - a rather inefficient and tedious process in which bigger stones are broken up, however, only simple machinery is required. The broken-up lumps of soil are then passed through a screen.

6.4   Pulverization of soil

The material is hit with great force so it disintegrates. The machinery required is complex but performs satisfactorily. At the delivery end, any large pieces left can be removed utilizing the screen.

6.5   Sieving

Soil contains various sizes of grain, from a very fine dust up to pieces that are still too large for use in block production. The oversized material should be removed by sieving, either using a built-in sieve, as with the pendulum crusher, or as a separate operation.

The simplest sieving device is a screen made from a wire mesh, nailed to a supporting wooden frame and inclined at approximately 45° to the ground. The material is thrown against the screen, fine material passes through and the coarse, oversized material runs down the front. Alternatively, the screen can be suspended horizontally from a tree or over a pit. The latter method is only suitable in the case where most material can pass through easily otherwise too much coarse material is collected, and the screen becomes blocked and needs frequent emptying.

6.6   Proportioning

Before starting production, tests should be performed to establish the right proportion of soil, stabilizer, and water for the production of good quality blocks. The proportions of these materials and water should then be used throughout the production process. To ensure uniformity in the compressed stabilized earth blocks produced, the weight or volume of each material used in the block-making process should be measured at the same physical state for subsequent batches of blocks. The volume of soil or stabilizer should ideally be measured in dry or slightly damp conditions. After establishing the exact proportion required of each material, it is advisable to build a measuring device for each material. The dimensions of each measuring box should be such that their content, when full, is equivalent to the proportion which should be mixed with other materials measured in other gauge boxes. Alternatively, a simple gauge box may be used for all materials. In this case, the amount of material for the production of a given batch of blocks may be measured by filling and emptying the gauge box several times for each separate material. For example, a batch of blocks may require ten gauge boxes of soil for one gauge box of stabilizers.

Water may be measured in a small tank or container. It is advisable to mix enough materials to allow the block-making machine to operate for approximately one hour. Thus, the volume of the mixed material will depend on the hourly output of the block-making equipment.

6.7   Mixing

To produce good quality blocks, it is very important that mixing be as thorough as possible. Dry materials should be mixed first until they are a uniform color, then water is added and mixing continued until a homogeneous mix is obtained. Mixing can be performed by hand on a hard surface, with spades, hoes, or shovels. It is much better to add a little water at a time, sprinkled over the top of the mix from a watering can with a rose spray on the nozzle. The wet mix should be turned over many times with a spade or other suitable tool. A little more water may then be added, and the whole mixture turned over again. This process should be repeated until all the water has been mixed in.

When lime is used as a stabilizer, it is advisable to allow the mix to stand for a short while before molding starts to allow better moistening of soil particles with water. However, if cement is used for stabilization, it is advisable to use the mix as soon as possible because cement starts to hydrate immediately after it is wetted and delays will result in the production of poor quality blocks. For this reason, the quantity of cement-sand mix should not exceed what is needed for one hour of operation. Even so, the blocks produced at the end of one hour may be considerably weaker than those produced immediately after the mixing. A concrete mixer, even if available, will not be useful for mixing the wet soil, since the latter will tend to stick on the sides of the rotating drum. If machinery is to be used for mixing, it should have paddles or blades that move separately from the container. However, field experience shows that hand-mixing methods are often more satisfactory, more efficient, and cheaper than mechanical mixing, and are less likely to produce small balls of soil that are troublesome at the block molding stage.

6.8   Quantity of materials needed

Compressed stabilized earth building blocks are usually larger than traditional burnt bricks. Typical block size is 290 x 140 x 90mm. Its production will need about 7.5 to 8.5kg of materials depending on the compaction pressure. The exact amount of stabilizer necessary must be established for any particular project. The fraction of lime or cement usually varies between 5% - 8% by weight. Similarly, the optimum water content (OWC) for any particular soil must be determined experimentally. The moisture level varies widely with the nature of the soil. An approximate estimate of about 15% by weight is often assumed.

7.   Molding of stabilized mud blocks

7.1   Standards for block production

Many aspects should be taken into consideration before launching an operation to produce compressed stabilized earth building blocks:

- Amount and type of stabilizer required,
- Soil properties and its suitability for stabilization,
- Building standards and hence the quality of blocks required,
- Load-bearing requirements of construction i.e. single storey or more.

One of the purposes of this handbook is to make the reader aware of the problems associated with compressed earth blocks in the construction industry especially in developing countries where building standards have not yet been developed in the field of earth construction.

Generally, there is a wide variation of acceptable standards that vary according to local weather conditions. Blocks with wet compressive strengths in the range of 2.8 MN/m2 or more should be adequate for one and two-story buildings. Blocks of this type would probably not require external surface protection against adverse weather conditions. For one-storey buildings, blocks with a compressive strength in the order of 2.0 MN/m2 will probably be strong enough, but where rainfall is high an external treatment is necessary. Since the wet strength of a compressed stabilized earth block wall may be less than two-thirds of its dry strength. It should be remembered that all compressive strength tests should be carried out on samples that have been soaked in water for a minimum of 24 hours after the necessary curing period.

The final wet compressive strength of a compressed earth block depends not only on soil type, but also on the type and amount of stabilizer, the molding pressure, and the curing conditions. In 1998, standards for compressed earth blocks were ratified as African Regional Standards (ARS) under the auspices of the African Regional Organization for Standardization (ARSO) technical committee on building and civil engineering (ARSO/TC3) after having satisfied procedures for the approval of regional standards.

7.2   Testing soil before block production

For block production, the soil mix must be checked for each batch of blocks to attain the optimum moisture content (OMC).

Two simple field tests can be carried out. These are explained below:

- Take a handful from the soil mix for block production and squeeze it in the hand, the mix should ball together. When the hand is opened, the fingers should be reasonably dry and clean.

- Drop the ball sample onto a hard surface from a height of about one meter. If the sample:- completely shatters, this shows that it is not sufficiently moist,- squashes into a flattened ball or disc on impact with the hard surface, this implies too high a moisture content,- breaks into four or five major lumps, this shows that the moisture contents or the soil mix are close to the optimum moisture content (OMC).

To manufacture blocks of uniform size and density, special precautions must be taken to fill the mold with the same amount of mix for each compaction by using a small wooden box as a measuring device. To facilitate the development of the pressed blocks and to ensure good neat surfaces it is advisable to moisten the internal faces of the machine mold with a mold releasing agent (reject oil) which can be applied with either a rag, brush, or spray.

7.3   Curing

To achieve maximum strength, compressed stabilized earth blocks need a period of damp curing, where they are kept moist. This is a common requirement for all cementitious materials. What is important is that the moisture of the soil mix is retained within the body of the block for a few days. If the block is left exposed to hot dry weather conditions, the surface material will lose its moisture and the clay particles tend to shrink. This will cause surface cracks on the block faces. In practice, various methods are used to ensure proper curing. Such methods include the use of plastic bags, grass, leaves, etc. to prevent moisture from escaping.

After two or three days, depending, on the local temperatures, cement stabilized blocks complete their primary cure. They can then be removed from their protective cover and stacked in a pile. As the stack of blocks is built up, the top layer should always be wetted and covered, and the lower layer should be allowed to air-dry to achieve maximum strength. Alternatively, freshly molded blocks can be laid out in a single layer, on a non-absorbent surface, and covered with a sheet to prevent loss of moisture. The required duration of curing varies from soil to soil and, more significantly, which type of stabilizer is used. With cement stabilization, it is recommended to cure blocks for a minimum of three weeks. The curing period for lime stabilization should be at least four weeks. The compressed stabilized earth blocks should be fully cured and dry before being used for construction [3].

8.   Advantages of compressed stabilized earth blocks (CSEB)

There are many advantages of ompressed Stabilized Earth Blocks (CSEB). They are summarized as below:

8.1   A local material

Ideally, the production is made on the site itself or in the nearby area. Thus, it will save transportation, fuel, time, and money.

8.2   A bio-degradable material

Well-designed CSEB houses can withstand, with a minimum of maintenance, heavy rains, snowfall, or frost without being damaged. The strength and durability has been proven since half a century. But let's imagine a building fallen down and that a jungle grows on it: the bio-chemicals contained in the humus of the topsoil will destroy the soil-cement mix in 10 or 20 years and CSEB will come back to our mother earth!

8.3   Limiting deforestation

Firewood is not needed to produce CSEB. It will save the forests, which are being depleted quickly in the world, due to short view developments and the mismanagement of resources.

8.4   Management of resources

Each quarry should be planned for various utilizations: water harvesting pond, wastewater treatment, reservoirs, landscaping, etc. It is crucial to be aware of this point: very profitable if well managed, but disastrous if unplanned.

8.5   An adapted material

Being produced locally it is easily adapted to the various needs: technical, social, cultural habits.

8.6   A transferable technology

It is a simple technology requiring semi skills, easy to get. Simple villagers will be able to learn how to do it in few weeks. An efficient training center will transfer the technology in a week.

8.7   Job creation opportunity

CSEB allows unskilled and unemployed people to learn a skill, get a job, and rise in social values.

8.8   Market opportunity

According to the local context (materials, labor, equipment, etc.), the final price will vary, but in most cases, it will be cheaper than fired bricks.

8.9   Reducing imports

Produced locally by semi-skilled people, no need to import from far away from expensive materials or transport over long distances heavy and costly building materials.

8.10   Flexible production scale

Equipment for CSEB is available from manual to motorized tools ranging from village to semi-industry scale. The selection of the equipment is crucial.

8.11   Energy efficiency and eco-friendliness

Requiring only a little stabilizer the energy consumption in an m3 can be from 5 to 15 times less than an m^3 of fired bricks. The pollution emission will also be 2.4 to 7.8 times less than fired bricks.

8.12   Cost efficiency

Produced locally, with a natural resource and semi-skilled labor, almost without transport, it will be cost-effective. More or less according to each context and one's knowledge, if done properly, it will be easy to use the most adapted equipment for each case.

8.13   Social acceptance

Demonstrated, for long, CSEB can adapt itself to various needs: from poor income to well-off people or governments. Its quality, regularity, and style allow a wide range of final house products. To facilitate this acceptance, banish from your language "stabilized mud blocks", for speaking of CSEB as the latter reports R & D done for half a century when mud blocks referred, in the mind of most people, as a poor building material.

9.   Disadvantages of compressed stabilized earth blocks (CSEB)

There are some disadvantages of Compressed Stabilized Earth Blocks (CSEB). They are summarized as below:

- Proper soil identification is required or lacks soil.
- Unawareness of the need to manage resources.
- Ignorance of the basics for production & use.

- Wide spans, high & long building are difficult to do.
- Low technical performances compared to concrete.
- Untrained teams producing bad quality products.
- Over-stabilization through fear or ignorance, implying outrageous costs.
- Under-stabilization resulting in low-quality products.
- Bad quality or un-adapted production equipment.
- Low social acceptance due to counterexamples (By unskilled people, or bad soil & equipment).

Conclusion

We can conclude that firstly, identification tests are important because they allow defining characteristics of the earth, to situate them concerning the suitability criteria, and therefore orient about the choice of the stabilizer. The behavior of the blocks differs depending on the treatment and dosage incorporated. The compressive strengths in dry and wet conditions increase with the dosage of the binder. Mixing cement + lime yielded the best resistance. Cement stabilized blocks are less resistant to wet. The different formulations have determined the best treatment. It is the mixture of cement + lime which has proved the best suitable treatment, and this, from the point of view of strength and of durability. Better compaction could generally improve behavior blocks moisture because as previously mentioned, the press used was manual and required a lot of effort to compact the blocks at their maximum density. We finally conclude that by using available lands in the environment, which have not necessarily ideal properties for the construction, there may be an appropriate treatment that achieves fairly satisfactory results, provided it complies with the good rules of implementation and take into account the cost of stabilizing products.

References

[1]   Information on https://www.thehindu.com/features/homes-and-gardens/understanding-stabilised-mud-blocks/article6815579.ece

[2]   Information on https://theconstructor.org/geotechnical/soil-stabilization-methods-and-materials/9439/

[3]   Information on https://www.seminarsonly.com/Civil_Engineering/stabilized-mud-block.php

CHAPTER 7

Concrete Blocks

1.   Introduction

A concrete block is a construction product produced by casting concrete in a reusable mold or 'form' which is then cured in a controlled environment, transported to the construction site, and put or lifted into place. A precast concrete block is primarily used as a building material in the construction of walls. It is sometimes called a Concrete Masonry Unit (CMU). A concrete block is one of several precast concrete products used in construction. The term precast refers to the fact that the blocks are formed and hardened before they are brought to the job site. The production process for precast concrete blocks is performed on ground level, which helps with safety throughout a project. There is greater control of the quality of materials and workmanship in a precast plant rather than on a construction site. Financially, the forms used in a precast plant may be reused hundreds to thousands of times before they have to be replaced, which allows the cost of formwork per unit to be lower than for site-cast production [1]. Most concrete blocks have one or more hollow cavities, and their sides may be cast smooth or with a design. In use, concrete blocks are stacked one at a time and held together with fresh concrete mortar to form the desired length and height of the wall (Figure 1). Concrete block masonry has advantages over brick and stone masonry. Concrete blocks are manufactured in the required shape and sizes and these may be solid or hollow blocks. The common size of concrete blocks is 39cm x 19cm x (30cm or 20cm or 10cm) or 2 inch, 4 inch, 6 inch, 8 inch, 10 inch and 12-inch unit configurations.

Figure 1. A precast concrete block

2.   Historical background of concrete blocks

The concrete mortar was used by the Romans as early as 200 B.C. to bind shaped stones together in the construction of buildings. During the reign of the Roman emperor Caligula, in 37-41 A.D., small blocks of precast concrete were used as a construction material in the region around present-day Naples, Italy. Much of the concrete technology developed by the Romans was lost after the fall of the Roman Empire in the fifth century. It was not until 1824 that the English stonemason Joseph Aspdin developed Portland cement, which became one of the key components of modern concrete. The first hollow concrete block was designed in 1890 by Harmon S. Palmer in the United States. After 10 years of experimenting, Palmer patented the design in 1900. Palmer's blocks were 8 in (20.3 cm) by 10 in (25.4 cm) by 30 in (76.2 cm), and they were so heavy they had to be lifted into place with a small crane. By 1905, an estimated 1,500 companies were manufacturing concrete blocks in the United States. These early blocks were usually cast by hand, and the average output was about 10 blocks per person per hour. Today, concrete block manufacturing is a highly automated process that can produce up to 2,000 blocks per hour.

The concrete blocks were first used in the United States as a substitute for stone or wood in the building of homes. The earliest known example of a house built in this country entirely of concrete block was in 1837 on Staten Island, New York. The homes built of concrete blocks showed a creative use of common inexpensive materials made to look like the more expensive and traditional wood-framed stone masonry building. This new type of construction became a popular form of house building in the early 1900s through the 1920s. House styles, often referred to as "modern" at the time, ranged from Tudor to Foursquare, Colonial Revival to Bungalow. While many houses used the concrete blocks as the structure as well as the outer wall surface, other houses used stucco or other coatings over the block structure. Hundreds of thousands of these houses were built especially in the Midwestern states, probably because the raw materials needed to make concrete blocks were in abundant supply in sandbanks and gravel pits throughout this region. The concrete blocks were made with face designs to simulate stone textures: rock-faced, granite-faced, or rusticated. At first considered an experimental material, houses built of concrete blocks were advertised in many Portland cement manufacturers' catalogs as "fireproof, vermin proof, and weatherproof" and as an inexpensive replacement for the ever-scarcer supply of wood. Many other types of buildings such as garages, silos, and post offices were built and continue to be built today using this construction method because of these qualities [2].

3.   Raw materials for precast concrete blocks

The concrete commonly used to make concrete blocks is a mixture of powdered Portland cement, water, sand, and gravel. This produces a light gray block with a fine surface texture and a high compressive strength. A typical concrete block weighs around 17.2-19.5 kg. In general, the concrete mixture used for blocks has a higher percentage of sand and a lower percentage of gravel and water than the concrete mixtures used for general construction purposes. This produces a very dry, stiff mixture that holds its shape when it is removed from the block mold.

If granulated coal or volcanic cinders are used instead of sand and gravel, the resulting block is commonly called a cinder block. This produces a dark gray block with a medium-to-coarse surface texture, good strength, good sound-deadening properties, and a higher thermal insulating value than a concrete block. A typical cinder block weighs between 11.8-15.0 kg.

Lightweight concrete blocks are made by replacing the sand and gravel with expanded clay, shale, or slate. Expanded clay, shale, and slate are produced by crushing the raw materials and heating them to about 2000°F (1093°C). At this temperature, the material bloats, or puffs up, because of the rapid generation of gases caused by the combustion of small quantities of organic material trapped inside. A typical lightweight block weighs 10.0-12.7 kg and is used to build non-load-bearing walls and partitions. Expanded blast furnace slags, as well as natural volcanic materials such as pumice and scoria, are also used to make lightweight blocks. In addition to the basic components, the concrete mixture used to make blocks may also contain various chemicals, called admixtures, to alter curing time, increase compressive strength, or improve workability. The mixture may have pigments added to give the blocks a uniform color throughout, or the surface of the blocks may be coated with a baked-on glaze to give a decorative effect or to provide protection against chemical attack. The glazes are usually made with a thermosetting resinous binder, silica sand, and color pigments [3].

4.   Design of precast concrete blocks

The shapes and sizes of the most common concrete blocks have been standardized to ensure uniform building construction. The most common block size in the United States is referred to as an 8-by-8-by-16 block, with the nominal measurements of 8 in (20.3 cm) high by 8 in (20.3 cm) deep by 16 in (40.6 cm) wide. This nominal measurement includes room for a bead of mortar, and the block itself measures 7.63 in (19.4 cm) high by 7.63 in (19.4 cm) deep by 15.63 in (38.8 cm) wide.

Many progressive block manufacturers offer variations on the basic block to achieve unique visual effects or to provide desirable structural features for specialized applications. For example, one manufacturer offers a block specifically designed to resist water leakage through exterior walls. The block incorporates a water repellent admixture to reduce the concrete's absorption and permeability, a beveled upper edge to shed water away from the horizontal mortar joint, and a series of internal grooves and channels to direct the flow of any crack-induced leakage away from the interior surface. Another block design, called a split-faced block, includes a rough, stone-like texture on one face of the block instead of a smooth face. This gives the block the architectural appearance of a cut and dressed stone. When manufacturers design a new block, they must consider not only the desired shape but also the manufacturing process required to make that shape. Shapes that require complex molds or additional steps in the molding process may slow production and result in increased costs. In some cases, these increased costs may offset the benefits of the new design and make the block too expensive.

5.   Concrete block construction method

5.1   Materials used in making concrete blocks

Many types of aggregates have been used with success for making concrete blocks. These include crushed stone, gravel, sand coral, volcanic cinders, slag, foamed slag, furnace clinker, etc. In humid areas, cement must be carefully stored to prevent its deterioration through premature hydration. Water-free from impurities such as oils, acids, organic matter must be used. Where sulphate-bearing water is liable to attack underground concrete work, it is probably advisable to use stone or brick for foundations together with a sulphate resisting Portland cement mortar.

5.2   Mixes

The mix will vary according to the type of aggregate used, but it should not be richer than one part (by volume) of cement to six parts of mixed fine and coarse aggregates. For making blocks in which no coarse aggregate is used, a mix consisting of one part of cement to six or seven parts of well-graded sand is satisfactory; some users employ mixes up to one cement to eight or nine sand depending on the end-use of the blocks. For dense blocks, aggregates should be well-graded to ensure that the small particles occupy the spaces between the larger ones and leave a minimum of voids. Well-graded sand will produce a much denser block of greater strength and lower moisture movement (although of higher thermal conductivity), than a block made from poorly graded aggregate and sand.

5.3   Mixing

As much water should be added to the mix as will produce water sheen on the surface of the block, and still not cause the block to slump. The length of time in mixing is important.

5.4   Mold

Molds shall be fabricated using mild steel plates and mild steel angles for stiffening the plates. The mold should be either fixed type or (box with four side walls fixed at corners and top and bottom open) or split type. Split type may be either individual or gang mold. Where the compaction of the concrete is done manually, the mold may be either fixed type or split type. When the composition of the compaction of the blocks is done with a surface vibrator, the mold shall be only split type (individual or gang mold). Demolding shall be done 5 to 10 minutes after compaction. In the case of fixed type mold, it shall be pulled up with one side handle while pressing down the blocks with the place at the top with the thumb. In case of split mold, the sides shall be removed first and the partition plates (gang mold) shall be pulled up subsequently. After demolding, the blocks shall be protected until they are sufficiently hardened to permit handling without damage.

5.5   Proportion

The normal proportion of the mix shall be as specified. To attain maximum strength, the water-cement ratio and workability of the mix should or controlled and proper compaction of concrete in the mold ensures age.

5.6   Casting

After mixing, the concrete shall be placed in the molds immediately, being carried in suitable vessels or by a chute. The concrete should be mixed in a concrete mixer and the water-cement ratio should be minimum. The casting of blocks should be done, preferably by using a small immersion vibrator or table vibrator. The complete operation being the work of a few seconds only for a practiced operator, provided he is kept supplied with the mix, can be expected to produce approx. 90 solid blocks of size 18"x9"x4-1/2" per hour. As the mold box is only filled up once, the same amount of concrete is used for each slab of anyone thickness – a great help when working up estimates. A huge daily output could be realized when five or six machines are run in co-operation with a motor – mixer.

5.7   Curing

The blocks hardened shall then be cured in a curing water tank or a curing yard and shall be kept continuously moist for at least 14 days. The blocks should be sheltered from sun and drying winds. After 24 hours they should be watered and kept damp. Once molded blocks have sufficiently hardened to permit removal of the supporting wooden pallet they may be carefully turned on side or edge and the pallet removed, the pallet oiled and reused. Keep blocks damp for several days to permit the cement to hydrate completely. The longer the curing time the better is the strength. The blocks should thereafter be completely dried before placing in the wall.

5.8   Drying

After curing, the blocks shall be dried for a period of two to four weeks depending upon weather before being used on the work. The blocks shall be allowed to complete their initial shrinkage before they are laid in a wall [2].

6.   Concrete block specifications

6.1   Dimensions

Concrete masonry building units shall be made in sizes and shapes to fit different construction needs. They include a stretcher, corner, double corner or pier, jamb, header, jamb, bullnose, and partition block, and concrete floor units. Concrete block hollow (open or closed) or solid shall be referred to by its nominal dimensions. The maximum variation in the dimensions shall not be + 1.5 mm for height and breadth and + 300mm length (Figure 2). The size other than those mentioned above may also be used by mutual agreement between the purchaser and manufacturers/suppliers. The nominal dimensions of the concrete block shall be, as follows:

Length: 400,500 or 600mm

Height: 200 or 100mm

Width: 50, 75, 100, 200, 250 or 300mm

In addition, the block shall be manufactured in half lengths of 200, 250, or 300mm to correspond to the full-lengths.

Figure 2. An architectural precast concrete block

6.2   Cavities

The total width of the cavity in a block right angle to the face of the block as laid in-wall (i.e. the bedding surface will be at right angles to the face of the block shall not exceed 65% of the total breadth of the block.

6.3   Shell thickness

The shell thickness of blocks shall not be less than 40mm for sizes A and B and 20mm for other sizes.

6.4   Joints

The end of the blocks, which form the vertical joints, may be a plain rectangle and grooved or double grooved.

6.5   Aggregates

All aggregates shall pass through I.S. sieve 1285mm and not more than 12 percent shall pass through I.S. sieve 300 microns. In addition, at least 15% shall be retained on I.S. sieve 10mm and 40 percent on I.S. sieve 4.75mm. The fineness modular of the combined aggregates may be between 3.6 to 40.

6.6   Density

The block density hollow concrete block shall not be richer than 1600 kgs. Per cubic meter of gross volume

6.7   Crushing strength

The average crushing strength of eight blocks that are immersed in the water maintaining a temperature of $27°C + 2°C$ for 24 hours bedded with cement sand mortar shall be not less than 50 kg per square cm of gross area.

6.8   Drying shrinkage

The drying shrinkage of an average of three blocks when unrestrained shall not exceed 0.004 percent. The moisture content (average of 3 blocks) of the dried blocks on being immersed in water shall not exceed 0.03 %.

6.9   Tolerances

The maximum variation in the length of the units shall not be more than +5mm and the maximum variation n height and width of the unit, not more than +3mm.

6.10   Density

The hollow blocks shall be provided cavities in such a way to ensure the maximum block density of 1600 kg/m^3. The block density is the density calculated by dividing the mass of a block by the overall volume including holes or cavities.

7.   Types of concrete block or concrete masonry units

Concrete blocks are produced in a large variety of shapes and sizes, solid, cellular or hollow, dense or lightweight, air-cured or steam-cured, loadbearing or non-loadbearing, and can be produced manually or with the help of machines. The block sizes are usually referred to by their nominal dimensions, which are the actual block length, width, and height plus 10 mm of mortar bed thickness added to each dimension. These are normally based on the modular coordination of design with the 10cm module as its basic unit. The most commonly used concrete blocks are the stretcher blocks with a nominal length of 40 cm (half blocks: 20 cm) nominal height of 20 cm, and nominal widths of 8, 10, 15, and 20 cm. In addition, a wide variety of non-modular blocks and special shapes are available, such as a corner, jamb, lintel, pilaster, and interlocking blocks, to name only a few.

The solid blocks have no cavities, or - according to US standards - have voids amounting to not more than 25 % of the gross cross-sectional area. Thinner blocks of less than 75 mm (3 inches) width are essentially solid, because of the difficulty of forming cavities. The cellular blocks have one or more voids with the one-bed face closed, and are laid with this 'blind end' upwards, preventing wastage of bedding mortar, which would otherwise drop into the cavities. The hollow blocks are the most common types of concrete blocks, having one or more holes that are open at both sides. The total void area can amount to 50 % of the gross cross-sectional area, and - according to British Standards - the external wall thickness must be at least 15 mm or 1.75 x nominal maximum size of aggregate, whichever is greater. The use of concrete hollow blocks has several advantages:

- They can be made larger than solid blocks, and if the lightweight aggregate is used, can be very light, without forfeiting much of their load-bearing capacity;
- They require far less mortar than solid blocks (because of the cavities and less proportion of joints, due to large size), and construction of walls is easier and quicker;
- The voids can be filled with steel bars and concrete, achieving high seismic resistance;

- The air-space provides good thermal insulation, which is of advantage in most climatic regions, except warm-humid zones; if desirable, the cavities can also be filled with thermal insulation material;

- The cavities can be used as ducts for electrical installation and plumbing.

- Dense concretes are normal concretes with a density exceeding 2000 kg/m3, while the densities of lightweight concrete can be as low as 160 kg/m'. The former is produced with well-graded aggregates (with a large number of fines to fill all voids) and full compaction, while the latter comprises lightweight aggregates and/or a high proportion of single-sized particles of coarse aggregate (no-fees concrete) in a lean mix, which is not fully compacted, or comprise a sand-cement mix with a foaming agent to aerate the mixture. Lightweight concrete is generally used for concrete blocks, provided that the ingredients are available and the strengths obtained are acceptable.

- Air curing is the standard procedure for the strength development of concrete, by which the concrete is kept wet for at least 7 days and then allowed to dry at ambient temperature. With steam curing, by which the concrete is exposed to low or high-pressure steam (in autoclaves), high early strengths can be achieved (with autoclaving the 28-day strength of air-cured concrete can be obtained in 24 hours). However, in developing countries, steam curing is unlikely to be implemented, because of its high cost and sophistication.

- The definition of loadbearing and non-loadbearing blocks is fairly complex and depends not only on the compressive strengths of the blocks but also on the ratio of their height to thickness, their density, and the proportion of voids.

- Manual block production is the cheapest but most laborious method, and the blocks are not likely to attain the superior qualities that are achieved by the far more expensive mechanized production.

Depending upon the structure, shape, size, and manufacturing processes, the concrete blocks are mainly classified into two types [4]. They are summarized as below:

- Solid concrete blocks
- Hollow concrete blocks

7.1 Solid concrete blocks

Solid concrete blocks are commonly used, which are heavy and manufactured from the dense aggregate. They are very strong and provide good stability to the structures. So for large work of masonry like for load-bearing walls these solid blocks are preferable (Figure

3). They are available in large sizes compared to bricks. So, it takes less time to construct concrete masonry than brick masonry.

Figure 3. A solid concrete block

7.2 Hollow concrete blocks

Hollow concrete blocks contain void areas greater than 25% of gross area. The solid area of hollow bricks should be more than 50%. The hollow part may be divided into several components based on our requirements. They are manufactured from lightweight aggregates. They are lightweight blocks and easy to install.

8. Types of hollow concrete blocks

- Stretcher block
- Corner block
- Pillar block
- Jamb block
- Partition block
- Lintel block
- Frogged brick block
- Bullnose block

8.1 Concrete stretcher blocks

Concrete stretcher blocks are used to join the corner in the masonry. Stretcher blocks are widely used concrete hollow blocks in construction. They are laid with their length parallel to the face of the wall (Figure 4).

Figure 4. A concrete stretcher block

8.2 Concrete corner blocks

The concrete corner blocks are used at the ends or corners of masonry. The ends may be window or door openings etc. they are arranged in a manner that their plane end visible to the outside and the other end is locked with the stretcher block (Figure 5).

Figure 5. A concrete corner block

8.3 Concrete pillar blocks

The concrete pillar block is also called a double corner block. Generally, these are used when two ends of the corner are visible (Figure 6). In the case of piers or pillars, these blocks are widely used.

Figure 6. A concrete pillar block

8.4   Jamb concrete blocks

Jamb blocks are used when there is an elaborated window opening in the wall. They are connected to a stretcher and corner blocks (Figure 7). For the provision of double-hung windows, jamb blocks are very useful to provide space for the casing members of the window.

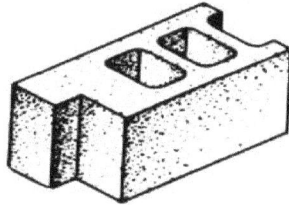

Figure 7. A jamb concrete block

8.5   Partition concrete blocks

Partition concrete blocks are generally used to build partition walls. Partition blocks have a larger height than their breadth. The hollow part is divided into two to three components in the case of partition blocks (Figure 8).

Figure 8. A partition concrete block

8.6   Lintel blocks

Lintel block or beam block is used for the provision of beam or lintel beam. Lintel beam is generally provided on the top portion of doors and windows, which bears the load coming from the top. Concrete lintel blocks have a deep groove along the length of the block as shown in the figure. After placing the blocks, this groove is filled with concrete along with reinforcement (Figure 9).

Figure 9. A concrete lintel block

8.7 Frogged brick blocks

Frogged brick block contains a frog on its top along with a header and stretcher like a frogged brick. This frog will help the block to hold the mortar and to develop a strong bond with the top laying block (Figure 10).

Figure 10. A frogged concrete block

8.8 Bullnose concrete block

Bullnose blocks are similar to corner blocks. Their duties also the same but when we want rounded edges at the corner bullnose bricks are preferred (Figure 11).

Figure 11. A bullnose concrete block

9.   Size of concrete blocks

Table1. Different sizes of concrete blocks

S. No.	Type of Blocks	Size in MM	Purpose
1.	Solid Blocks	100X200X400 200X200X400	Foundation
2.	Closed Cavity Blocks	75X200X400 100X200X400 150X200X400 200X200X200	Load-bearing External works Partition walls
3.	Corner Column Blocks	200X200X400	Corners
4.	Roofing Blocks	410X250X140 530X250X140	Roofs
5.	Bend (U) Blocks	100X200X200 200X200X200	R.C.C. Bend

10.   Building with concrete blocks

10.1   Design

- To minimize the need for cutting concrete blocks, all horizontal dimensions of walls should be multiples of nominal half blocks (most commonly 20 cm) and all vertical dimensions should be multiples of nominal full-heights (20 cm).

- To minimize the risk of cracking, the lengths of individual wall sections should not be greater than one-and-a-half times the height.

- Hollow blocks should be specified when good thermal insulation is required. These blocks are also useful when additional structural stability is needed, eg in earthquake areas, because the cavities align vertically and can be filled with reinforcing steel and concrete.

- Blocks with a rough surface (open textured), as in the case of most lightweight blocks, are advantageous, because they provide a better key for bedding mortar and applied finishes and have a less capillary attraction for water, and dry out more quickly after rains.

10.2 Construction

- Concrete blocks must be dried out thoroughly before use, otherwise drying will continue after building the wall and shrinkage cracks may develop.

- Only dry blocks should be used and they should not be wetted before laying. Instead, the preparation of the mortar must take into consideration that the blocks absorb some of the water.

- Mortars used for bedding should not be too rich in cement. Cement: hydrated lime: sand mixes of 1: 2: 9 or 1: 1: 6 have high water retention and good workability. The strength of the mortar mustn't exceed that of the blocks, so that the joints can absorb a limited amount of movement, preventing the blocks from cracking [5].

11. Construction of precast concrete block masonry

For single-storeyed buildings, the hollows of blocks in foundation and basement masonry shall be filled up with sand and only the top foundation course shall be of solid blocks. But for two or more storeyed buildings, solid concrete blocks shall be used in foundation courses, plinth, and basement walls, unless otherwise indicated. If hollow blocks are used, their hollows shall be filled up with cement concrete 1:3:6 using 12.5 mm nominal size aggregates.

11.1   Wetting of blocks

Blocks need not be wetted before or during laying in the walls. In case the climate conditions so require, the top and the sides of blocks may only be slightly moistened to prevent absorption of water from the mortar and ensure the development of the required bond with the mortar.

11.2   Laying

Blocks shall be laid in mortar, as indicated and thoroughly bedded in mortar, spread over the entire top surface of the previous course of blocks to a uniform layer of not less than 10 mm and not more than 12mm in thickness. All courses shall be laid truly horizontal and vertical joints made truly vertical Blocks shall break joints with those above and below for not less than a quarter of their length. Precast half-length closer and not cut from full-size blocks shall be used. For the battered face, bedding shall be at a right angle to the face unless otherwise directed.

11.3   Provision for door and window frames

A course of solid concrete block masonry shall be provided under door and window openings (or a 10 cm thick precast concrete sill block under windows). The solid shall extend for at least 20 cm beyond the opening on either side. For jambs, very large doors, and windows either solid units are used, or the hollows shall be filled in with concrete of mix 1:3:6 using 12.5 mm nominal size aggregates.

11.4   Intersecting Walls

When two walls meet or intersect and the course is to be laid up at the same time, a true masonry bond between at least 50% of the units at the intersecting is necessary. When such intersecting walls are laid up separately, pockets with 20mm maximum vertical spacing shall be left in the first wall laid. The corresponding course of the second wall shall be built into these pockets.

11.5   Provisions for roof

The course immediately below the roof slab shall be built with solid blocks. The top of the roofing course shall be finished smooth with a layer of cement and coarse sand mortar 1:3, 10 mm thick and covered with a thick coat of whitewash or crude oil, to ensure free movement of the slab.

11.6   Piers

The top course of the block in the pier shall be built in solid blocks. Hollow concrete blocks shall not be used for isolated piers unless their hollows are specified to be filled with cement concrete. Fixtures, fitting, etc. shall be built into the masonry in cement and carse sand mortar 1:3 while laying the blocks where possible. Hold fasts shall be built into the joints of the masonry during laying. Holes, chases, sleeves, openings, etc. of the required size and shape shall be formed in the masonry with special blocks while laying, for fixing pipes, service lines, the passage of water, etc. After service lines, pipes, etc. are fixed, voids left, if any, shall be filled up with cement concrete 1:3:6 (1 cement :3 coarse sand:6 stone aggregate 2 mm nominal size) and neatly finished.

11.7   Finishes

Rendering shall not be done to the walls when walls are wet. Joints for plastering or pointing as specified shall be raked to a depth of 12 mm. joints on internal faces unless otherwise indicated, shall be raked for plastering. If the internal faces of masonry are not to be plastered the joints shall be finished flush as the work proceeds or pointed flush where so indicated.

12.   Advantages of concrete blocks

There are many advantages of using concrete blocks [5]. They are summarized as below:

- The economy in design of sub-structure due to reduction of loads
- Saving in mortar for laying of blocks as compared to ordinary brickwork. Saving in mortar for plasterwork. Uniform Plaster thickness of 12 mm can be maintained due

to precision of the size of blocks as compared to brickwork where plaster thickness of average of 20 mm is required to produce uniform and even plastered surface due to variations in the sizes of bricks. Insulation of walls is achieved due to the cavity, which provides energy saving for all times. Similarly, hollowness results in sound insulation.

- Paint on finished walls can be applied due to cavity, which provides energy saving for all times. Similarly, hollowness results in sound insulation.

- No problem with the appearance of salts. Hence great savings in the maintenance of final finishes to the walls.

- The laying of Blocks is much quicker as compared to brickwork hence saving time.

13. Comparison of hollow block and brick masonry

The comparison of materials used for 100 sqm / 100 sqft of 200 mm and 100 mm thick hollow block walls with 9 inches and 4.5-inch brick wall respectively, is shown in Table 2.

Table 2. Comparative analysis of hollow block and brick masonry

Width of Wall	Units	200 mm	100 mm	215-225 mm	107-114 mm
Quantity of Hollow Blocks/Bricks required for 100 sqm of Surface Area	Nos.	1161	1161	11365-12120	5926-5435
Quantity of Hollow Blocks/Bricks required for 100 sqft of surface area	Nos.	108	108	1127-1056	551-505
Mortar for Laying Blocks/Bricks per 100 sq-m	cu-m	1.394	0.679	4.63-4.85	1.80-1.87
Mortar for Laying Blocks/Bricks per 100 sq-ft.	cu-ft	4.571	2.286	15.11-15.93	5.92-6.15
Mortar for plastering both sides of Brick Masonry for 12 mm/20 mm per 100 sq-m	cu-m	2.400	2.400	12.500	12.500
Mortar for plastering both sides of Brick Masonry for 12 mm/20 mm per 100 sq-ft	cu-ft	8.333	8.333	2.500	12.500

Conclusion

A precast concrete block is a concrete member that is cast and cured at a location other than its final designated location. The simple concrete block will continue to evolve as architects and block manufacturers develop new shapes and sizes. These new blocks promise to make building construction faster and less expensive, as well as result in structures that are more durable and energy-efficient. Some of the possible block designs for the future include the biaxial block, which has cavities running horizontally as well as vertically to allow access for plumbing and electrical conduits; the stacked siding block, which consists of three sections that form both interior and exterior walls; and the heat soak block, which stores heat to cool the interior rooms in summer and heat them in winter.

References

[1]   Allen, E. A. Fundamentals of building construction materials and methods. Hoboken, NJ: John Wiley & Sons, Inc. 2009.

[2]   Techno-Economic Feasibility Report on Concrete Hollow and Solid Block, Building Materials and Technology Promotion Council, New Delhi, India, 2010.

[3]   Information on http://www.theconstructor.org/concrete/precast-concrete-blocks/7263/

[4]   Information on https://theconstructor.org/building/types-concrete-blocks-masonry-units/12752/

[5]   Techno-Economic Feasibility Report on Concrete Hollow and Solid Block, Building Materials and Technology Promotion Council, New Delhi, India, 2010.

CHAPTER 8

Reinforced Hollow Concrete Block Masonry

1.   Introduction

The Reinforced Hollow Concrete Block Masonry (RHCBM) is an alternative cost-effective walling replacing R.C.C. framed structure. RHCBM can be used as load-bearing walls even for multi-storied buildings (up to 5 stories) being constructed with hollow blocks. It is lightweight. With proper joints and reinforcement, the structure constructed with this wall can also be designed to be earthquake-resistant. With this load-bearing system, the traditional beams and columns can be eliminated. In this technology, the concrete blocks having two holes and block size 200mm x 200mm x 400mm are laid with staggered joints. Vertical Reinforcement is placed through these holes as per design. Only those holes which carry the reinforcement are grouted with concrete, the others remain hollow (thus the building is lightweight.)

The concept of reinforced masonry utilizes the floors and roof as diaphragms acting as horizontal flanged girders to distribute lateral loads to walls. The walls provide horizontal shear resistance needed, in addition to carrying the normal vertical dead and live loads. This type of structure is defined as a Box system or Bearing wall System. For a building to be earthquake-resistant, the following three factors are to be taken into account:
- The building should be lightweight.
- Structure behavior should be of box type load-bearing (i.e. the walls to act as shear walls and roof as the diaphragm).
- Openings and corners are stiffened with horizontal and vertical reinforcement.

Thus, the 200mm thick RHCBM wall acts as a shear wall with a slab as the diaphragm. As the blocks are load-bearing, the load is assumed to be evenly distributed on the footing. Continuity avoids any differential settlement also.

2.   Reinforced hollow concrete block wall

The load-bearing structures whether of brick walls or hollow concrete block walls are safe, economical for normal dead and live loads as these loads are vertical and induce mostly compressive stresses. During earthquakes, cyclones, the horizontal (lateral) load becomes prominent and induces high tensile and shear stresses in the masonry walls. To make the structure earthquake and cyclone-resistant coupled with the economy, the walls are reinforced. By proper detailing of the reinforcement in the walls, it is possible to structurally integrate the walls to roofs and walls to walls and also ensure that there is shear wall action in every masonry element and the diaphragm action in the roof/floor. The behavior of the structure is like a 3D-Box. Reinforcing a brick wall has some practical difficulties, that it requires some shuttering for reinforcing brick masonry. It breaks the continuity and integrity of masonry and that the reinforcement may get corroded. Reinforcing hollow block masonry is much simpler and effective, because of the hollow spaces of the hollow blocks. The reinforcement can be provided vertically in the hollow spaces, without breaking the continuity of the masonry. The horizontal reinforcement is provided through the U-Channel block course. The reinforced hollow concrete block masonry (RHCBM) technique effectively utilizes the structural hollow concrete blocks for an efficient structural system, rather than its use only as infill in the RC framed structures. In principle, the reinforcement helps to develop the flexural tensile resilience and ductility needed for high-rise load-bearing structures [1].

3. Structural hollow concrete blocks

There is a difference between the hollow concrete blocks and the structural masonry blocks in terms of their dimensional consistency, quality, and strength. Normally, conventional hollow blocks are manufactured and used in place of bricks as infill in the RC framed system where bricks of good quality are not available; for a better quality of walling, fewer joints, and dimensional stability. But in such walls, their use is limited as infill walls only due to the lower strength of these blocks [2]. The structural hollow concrete blocks are of high strength and are used as load-bearing with adequate ductile detailing (Figure 1).

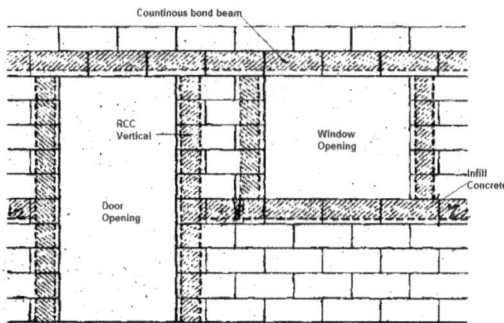

Figure 1. Structural bands for improved stability in RHCB.

4.   Structural features

- The block is vertically reinforced with steel reinforcement and grouting concrete of minimum M-20 filled at regular intervals, through the common vertical cores of hollow concrete blocks.

- Each block is horizontally reinforced with steel reinforcement and grouting concrete-filled at plinth level, lintel level, and roof level as, continuous R.C. band using U channel-shaped blocks in the masonry course.

- The reinforcement and grouting concrete is done without any shuttering.

- There is no scope for corrosion or reinforcement.

- As the reinforcement bars in both vertical and horizontal directions are continued into the roof slab and the walls respectively, the structural integrity is achieved in all three directions.

- In this technique, practically every wall and slab behave as a shear wall and a diaphragm respectively, thus increase in structural safety and stability. Due to the uniform distribution of reinforcement in both vertical and horizontal directions, increases the tensile resistance and ductile behavior of walls are increased, and therefore, the structure offers better resistance under dynamic loading.

- The technique is economical as it is a combined load-bearing and a shear wall structural system. Figure 2 shows a typical 3D cut-through view showing various features of RHCB Masonry.

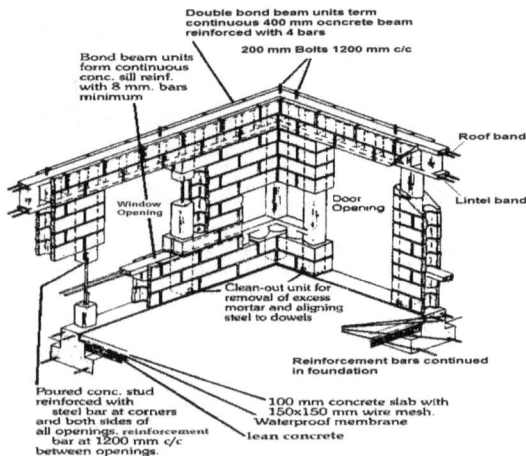

Figure 2. Typical sectional view showing various features of RHCB masonry

Materials Research Forum LLC
https://doi.org/10.21741/9781644901519

5.   Precaution while laying RHCB masonry

The following precautions should be kept in mind while laying RHCB Masonry:

- The technology requires effective supervision and care during the tying of reinforcement, placement, grouting, and compaction.

- The hollow concrete blocks should be of good quality concrete (Min. M-15).

- The grout concrete should not be less than M-20.

- Care should be taken to place the vertical reinforcement in the center of the hollow space of the blocks.

- For overlapping, tie the vertical bar first and then grout the hollow cores.

- There is a practical difficulty in the placement of hollow blocks over the vertical bars.

- Use 12mm and down aggregate, as the grout concrete has to fill the narrow hollow cores.

- Use plasticizers for grout concrete, as mechanical vibration cannot be done.

- For horizontal reinforcement use U-Shaped blocks.

- Continue the vertical bars into the next floor after lapping. Provide extra bent-up bars and bend into the slab to structurally integrate the wall to the roof (Figure 3).

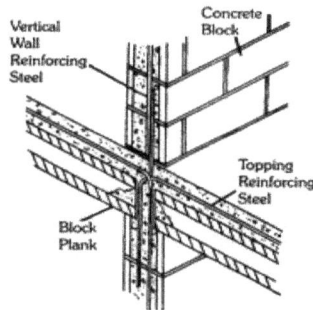

Figure 3. Wall and slab junction

- There is a practical difficulty in the placement of hollow blocks over the vertical bars which require lifting of hollow blocks to the height of the vertical reinforcement end (Figure 4).

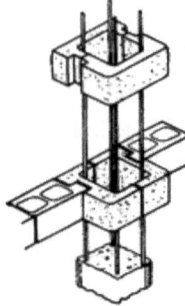

Figure 4. Placement of hollow blocks over vertical bars

- For convenience, the blocks with split cut in one face can be manufactured and they can be placed at ease (Figure 5).

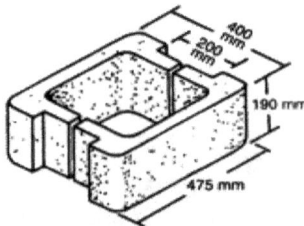

Figure 5. Split cut blocks

6. Design details

Figure 6 shows an example of two-bedroom twin flats marked as A and B in a 4 story residential block with load-bearing reinforced hollow concrete block technology. Figure 7 shows the longitudinal section of the common wall with no openings (marked as W7), showing vertical reinforcement from below the plinth beam. Figure 8 shows the longitudinal section of wall W-16 with Door opening (marked in plan) showing vertical reinforcement for the Ground floor. Figure 9 shows the same wall for the 2nd and 3rd floor with a reduction in diameter of vertical reinforcement. Figure 10 shows the longitudinal section of wall W-3 with door openings. Figure 11 shows the longitudinal section of external wall W-14, with window openings.

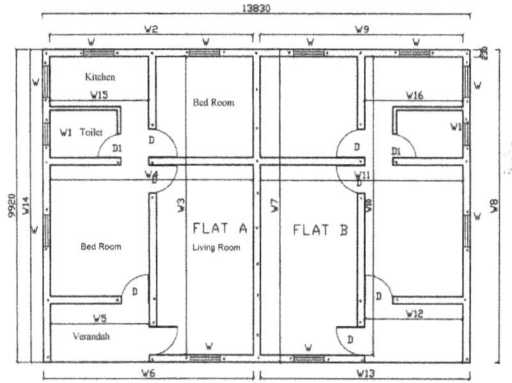

Figure 6. Typical floor plan showing wall panels for design analysis

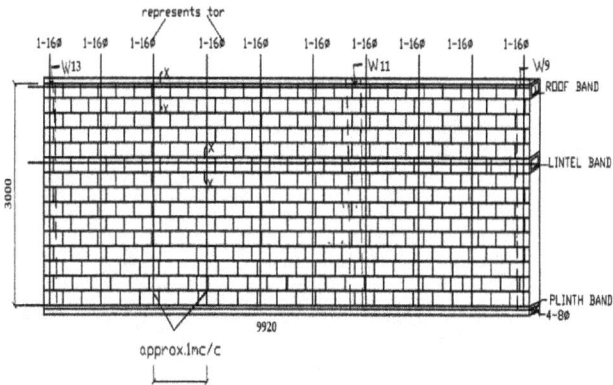

Figure 7. Reinforcement details for RHCB wall panel W7 on ground floor

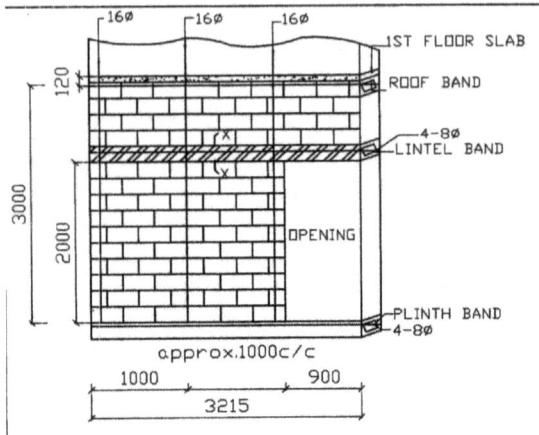

Figure 8. Reinforcement Details for RHCBM Wall Panel W16 in Ground and First Floor

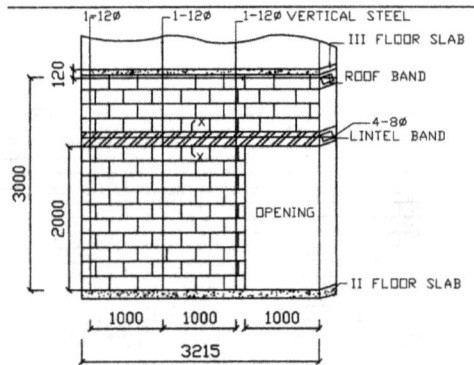

Figure 9. Reinforcement Details for RHCBM Wall Panel W16 in 2nd and 3rd Floor

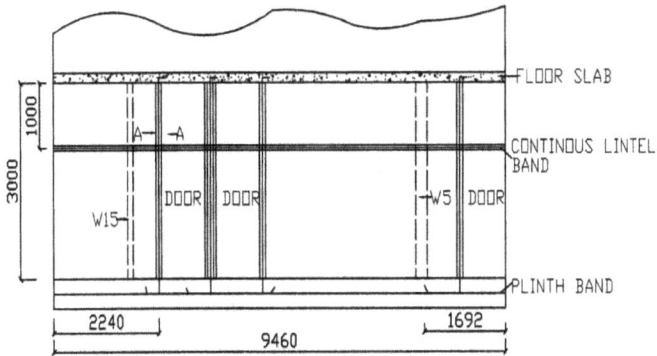

Figure 10. Sectional Elevation of Wall W3 at Ground Floor

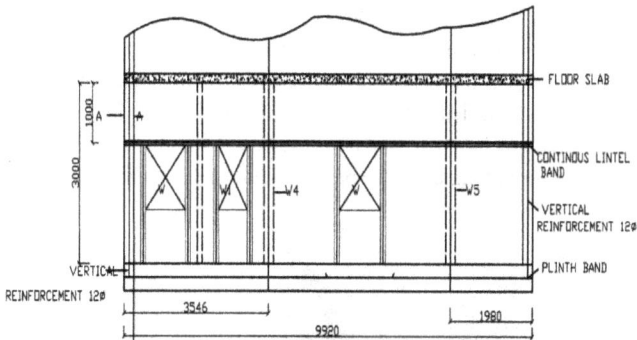

Figure 11. Sectional elevation of wall W14 at ground floor

References

[1]   P. K. Adlakha, New Building Materials, and Technologies, Vol. IV, Indian Building Congress, New Delhi, India, 2019.

[2]   G. N. Kamat, Reinforced concrete block masonry (RCB) as an earthquake-resistant structure, 2013, information on https://aceupdate.com/2013/10/17/green-buildings-by-innovative-masonry-method/

CHAPTER 9

Concrete Pavement Blocks

1.   Introduction

Block paving is one of the most popular flexible pavement surfacing options. It is made from various materials (concrete, clay, recycled plastic, etc.) but the most commonly used are clay and concrete. Paver blocks are used for exterior pavement applications. They are available in different sizes, shapes, colors, designs, thicknesses, textures, and patterns, which lend themselves to the creation of interesting designs in paved areas. In addition to breaking the monopoly of a flat paved area, colored concrete pavement blocks are used to permanently mark such things as parking bays, traffic lanes, and crossways. Interlocking pave blocks are high-strength pre-cast concrete elements designed for near medium and heavy-duty uses. Concrete pavement blocks offer an ecological solution to man's increasing concern for his environment, the marble of putting the moisture from rainfall back into the earth rather than let it run off as waste. The problem of continual repair to streets and sidewalks is simplified. Streets, for example, can be opened for utility line repairs and the pavers replaced to their original appearance without the usually ugly patch. The different types of interlocking concrete pavers blocks are shown in Figure 1.

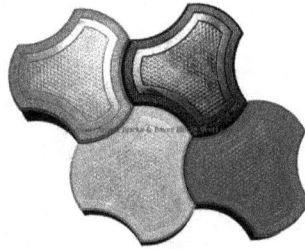

Figure 1. Different types of interlocking concrete pavers blocks

The concrete pavement blocks, interlocking, has the unique ability to transfer loads and stresses laterally employing an arching of bridging between units, spreading the load over a large area, reduces the stress thereby allowing heavier loads and traffic over sub-bases which normally would require heavily reinforced concrete. The process of manufacturing these rugged, shapely, and beautiful pieces of concrete moldings involves a unique vibration and hydraulic compaction. This imparts high compressive strength and durability apart from aesthetic beauty to the entire range of products. It's easy to add intriguing design, texture, and pattern to monotonous drives and walkways with the versatile paver available. Different systems meet a multitude of high-performance tasks and provide exciting design opportunities for landscape, architects, driveways, sidewalks, parking areas, pool decks, shopping alleys, canal linings, industrial floors, petrol pumps, loading decks, ramps, etc. are some of the applications. The concrete pavement blocks are ideal for medium traffic roads, sidewalks, garden paths, and public areas and it has excellent durability skid resistance, high strength, choice of colors, elegant appearance, and factory controlled quality. Concrete pavement block suits virtually all types of pavements and requires minimum maintenance. The range of concrete pavement blocks includes heavy-duty, medium duty, light duty, and decorative landscape pavers [1]. They are summarized in Table 1.

Table 1. Range of interlocking concrete pavers blocks

Light Usage	Medium Usage	Heavy Usage
Sidewalks/Walkways	Hotels-Driveway	Inland container/Depots
Garden Path	Restaurants (Sit-in-Areas)	Industrial Floors
Patios	Farm House-Driveway	Loading Docks/Ramps
Verandahs	Shopping Mall/Plazas	Ports/Maritime Terminals
Swimming Pool Decks	Amusement Parks	Petrol Pump/Service Station
Terraces/Rooftops	Holiday Resorts	Factory Compound
Pavements/Footpaths	Exhibition Grounds	Ware Houses
Jogging Track	Parking Lots	Bus Terminals
Bicycle Path	Embankments/Canal Lining	Street Escapes
Pedestrian Crosswalk	Railway Stations	Airfield runway

2.   Specifications of Grass Paver

The specifications of grass paver blocks are shown in Table 2.

Table 2. Specifications of grass paver concrete blocks

Blocks / Square Meter	4.15
Weight / Block	26 to 28 Kg
Void area on Block	40%
Surface Finish	Semi Coarse
Height of Block	100mm +/- 5mm
Length of Block	600mm +/-5mm
Breadth of Block	400 +/-5mm
Width of Shell (Between Grids)	50 +/-2mm
Distance Between Grids	90 +/-2mm

3.   Types of interlocking paver blocks

The interlocking paver block is a great material that is available in various shapes (I, Rectangular, square, Hexa, Cosmic, Trihex, Zig-Zag, and Cobble, etc.), colors (Red, Black, Yellow, Green, Blue and many other) and textures. Paver has a unique quality to interlock with each other due to the large availability of various shapes. The wide variety of design, color, and thickness of paver makes block paving suitable for both domestic and commercial applications including patio, driveway, sidewalk, pool deck, parking areas, etc. Figure 2 shows different types of interlocking concrete pavers.

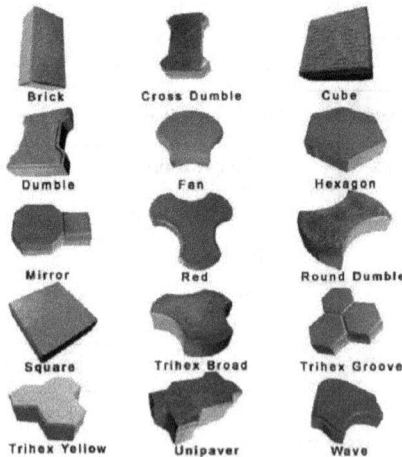

Figure 2. Different types of interlocking concrete pavers blocks

The different types of interlocking concrete pavers blocks are summarized as below:

3.1 Regular block (heavy duty)

High compressive strength and consistent quality controlled concrete blocks.

3.2 Irregular block (medium duty)

Extremely high-stress concrete block with unique interlocking shape

3.3 Grass paver block

An anesthetic look reduces the wastage of rainwater and absorbs it in the holes. Paver application varies widely and is practically unlimited difference systems satisfy the individual taste and meet multiple of high tasks.

3.4 Regular block (heavy duty)

Rectangular pavement blocks are functional, decorative, and simple in shape for heavy industrial block pavements, a block shape with straight sides is the appropriate choice for the following reasons:

- There is a better chance of a constant quality interlock around all sides of the block. The degree of interlock is less than a profiled block but its constancy is of greater importance.

- A rectangular block is less prone to stress concentration than a profiled block under horizontal loads, hence less chance of flexural tensile failure.

- A rectangular block pavement behaves better when subjected to traffic wave deformations than profiled blocks as horizontal forces will be transferred as line loads rather than points loads.

- From a practical point of view, rectangular blocks are easier to uplift and replace when services need to be installed under the pavement. This happens frequently at industrial sites.

3.5 Uni regular block (medium duty)

The irregular geometrical shape of the uni block provides an interlocking capacity significantly higher than conventional rectangular blocks. Uni blocks are suitable for almost all paving block applications from pedestrian sidewalks to heavy-weight vehicular traffic applications. The irregular block is an extremely high-stress concrete block with a unique interlocking shape. The irregular block is a multi-weaved paving block and has a unique "notched" design that allows each block to tightly interlock with the surrounding blocks, thus creating an incredibly strong, uniform surface with the beauty of blocks. Their design flexibility makes multi-weaved blocks the perfect modern option.

All the load acting on a natural pavement block outside its center of gravity causes a tipping motion. This is largely prevented by all-around "notching" of the irregular block. Every single block is not only secured against turning by four blocks in the neighboring row but by six surrounding blocks. It is, therefore, impossible for one single block to turn without the neighboring block would have to turn because of the "notching" of the block. The herringbone pattern provides the best possible lock effect because of the intensive clasping of every single block. The same applies to the six neighboring blocks and every other block in the area.

3.6   Grass paver block

The grass pavers create a green surface with the attractive appearance of grass combined with the strength of concrete. The grass paver has been designed to suit a hard strong driveway that permits greening and water percolation, provides a safe walking surface, and offers all types of slope protection.

4.   Manufacturing of pavement blocks

4.1   Selection of raw materials

Raw materials like sand and aggregate are carefully selected and tested in govt. approved laboratory before use in commercial production. Sieve analysis and grading check as required for the mix is done in our site laboratory regularly or the quality control of the material-grading.

4.2   Batching and mixing

Raw material cement, sand, aggregate, etc. are selected in the predetermined proportion (by weight) as per design mix. They are weighted on the weighing scale of the batching plant. Water is directly fed into the mixer and the mix is thoroughly homogeneously mixed in a pan mixer, thereby the green concrete is ready for processing. The above process of weighing and mixing is done in batching plants to ensure the consistency of mix proportion a greater degree of quality control.

4.3   Transit of wet concrete mix

The green mix is directly released from batching plant mixer into the block-making hopper by suitable placing the block-making machine hopper below the mixer drum of the batching of the plant.

Materials Research Forum LLC
https://doi.org/10.21741/9781644901519

4.4   Casting of the block

Block making machine consists of feed drawers vibrating table, mold (with cavities) temper head hydraulic power pack, control, etc. The concrete is fed into the mold cavities (while the mold rests on the vibrator table) once the cavities are filled with green concrete the blocks are pressed and compacted by vibrant compaction (simultaneously vibrating and applying high hydraulic pressure). This ensures that blocks are cast in the desired shape as that of mold cavities. Now the fresh blocks are released on the pallets and move forward on the conveyor system from where they are taken to the stacking area (Stacking, handling, etc. depends upon the area availability, site layout, and other local site conditions as considered suitable).

4.5   Curing

The fresh blocks after the initial setting are covered by jute rolls/and are continuously moistened by sprinkling water to retain the initial moisture and to reduce the heat of hydration for a period of a minimum of 10 days from the casting.

4.6   Testing and quality control

Matured blocks after they are fully cured are tested in our site laboratory for the crushing strength of the blocks as per the BS 6717 with the help of compression testing machine blocks. Once they are tested and passed as required are ready for dispatch to the concerned laying sites.

5.   Laying of pavement blocks

5.1   Leveling of sub-base

A good base with proper compaction shall be done before the sand bedding, allowing surcharge for compaction. The sand should be screened with guide rods to achieve a uniform thickness of 50mm on average [1].

5.2   Laying of blocks

Blocks are loosely laid manufacturing joint gaps of 2-4mm, alignment is controlled by tightening alignment strings and adjusting the gaps line correction, etc. by the skillful and trained workforce.

5.3   Compaction of surface

Compaction is done by vibratory roller for heavy/medium duty paving to ensure a uniform acceptable level.

5.4   Joint filling of blocks

Sweep dry, fine sand over the surface, making sure it fills all the gaps between the pavers.

5.5   Final compaction

Vibrate the area again to give a Strong interlocking bond. Sweep off the excess sand and your pavement is now completed and ready to use instantly.

6.   Characteristics features of pavement blocks

6.1   Durable

It has superior structural integrity, high strength, economical, and cheaper comparing other conventional methods.

6.2   Practical

It has low maintenance, easy to replace/change color, simplifies subsurface access, skid-resistant, and immediately useable.

6.3   Simple installation

It can be easily installed on a bed of sand/mortar. It is easy to install without using any heavy machinery or equipment. After installation, you can immediately use block paving for walking and can be open to the traffic. Mortar is not required for the installation of the paver.

6.4   Versatile

It has a variety of colors, shapes, designs, and textures.

6.5   Flexible

It easily blends with old and new surroundings, adapts to environmental fluctuations.

6.6   Attractive

It adds color, warmth, human scale, elegance, beauty to the built environment.

6.7   Reusable

It is easy to replace damaged and cracked blocks.

7. Advantages over conventional rigid concrete pavement blocks

There are many advantages of interlocking concrete pavement blocks over convention rigid concrete blocks [2]. They are summarized as below:

7.1 Better resistance to wear and tear

The compressive strength of the Interlock concrete block is very high (50 MPa) as compared to 15-20 MPa in the case of reinforced concrete, hence the wear and tear of the surface are less compared to conventional concrete floors. Block paving is very durable and is expected to last reliably for at least 20 years. Cracking is not developed when they are properly interlocking with each other. It is a great material for residential and commercial applications and long-lasting. Pavers can withstand any type of vehicle load and pedestrian traffic with low maintenance. Pavers are used for any type of traffic like low, medium, and heavy. It is also able to withstand a lot of pressure so that it can also be used in airports and docks.

7.2 Better thermal resistance

Interlock concrete blocks, being segmental and due to joint gaps, temperature stresses do not develop in interlock paving unlike the reinforced concrete paving, where temperature stress is an important factor influencing the life of the pavement.

7.3 Ready to use after laying

Interlock concrete block paving is ready to use soon after it is laid, no curing time (28 days in case of R.C.C.) is needed at the site.

7.4 Environmental friendly

Paver blocks are more environment-friendly as compared to the solid plain surface. Solid surface causes water and chemical runoff that may pollute natural water sources. Block paving will also prevent rainwater from gathering or flooding, allowing you to create a better drainage system on your patio or driveway. There are also available as interlocking permeable pavers, where vegetation is planted in each grid. These types of pavers allow the re-use of storm water (Surface water in abnormal quantity resulting from heavy falls of rain or snow.) back into the grounds, rather than back into the street.

7.5 High limiting deflection

The elastic deflection limit in rigid concrete paving is very small (2-4 mm) whereas the limit in interlock block paving is high (20-25 mm). This stops the pavement from cracking due to local soil settlement.

7.6   All-weather construction

Interlock concrete block paving can be laid in all weathers since there is no risk of work getting hampered, whereas in the case of concrete paving it is not possible to cast the concrete in wet and raining conditions.

7.7   Weather resistant

Pavers can be used in any climate (winter, summer, and monsoon). In colder climates, where snow is common, concrete pavers are ideal. The frost, oils, and most chemicals are not affecting the paver block, so this quality has made them a popular choice in heavy industrial areas. Also in a colder climate, electric or liquid snow melting systems (Snow melting system prevents the build-up of snow and ice on walkways, patios, and roadways.) can be installed under pavers. A snow melting system can reduce accidents as well as reduce slip hazards. It can reduce expenses related to removing snow and reduce the damaging effects of freeze-thaw cycles.

7.8   Easy repair and maintenance

The compressive strength of the block being very high the blocks are not prone to corresponding blocks can be easily replaced without losing time. Whereas in the case of rigid concrete paving repair is not only frequent (due to settlements, cracks, wear and tear, etc.) but costly, cumbersome also. It requires a lot of time (for construction and curing) thereby affecting the normal working. The pavers block requires low maintenance as compared to the plain concrete or asphalt pavement. The Paver block does not need polish on the top surface of the paver and it will not need to be repainted in the future. You can regularly wash your patio or driveway with simple soap and water, so it will help to keep it looking fresh and in bright condition for many years.

The most important advantage of interlocking paver block is, if one block is damaged or stained with oil, diesel, or petrol spillages, it can be easily removed and replaced with a new one. It is easy to repair and replace as compared to asphalt, and or concrete pavement and no heavy machinery is used for repairing or replacing the block. No skilled labor is required to install these paver blocks.

7.9   Safety

Pavers are very safe pavement material for outdoor applications. When the top surface of paver is oily due to the passing of vehicles over them, it is very safe to walk and drive a vehicle over them. It is also safe when the top surface of the interlocking paver block is wet or in monsoon season, there is no problem for walking or no tension of slipping over them. Interlocking pavers are slip-resistant and skid-resistant.

7.10   Facility for post construction underground works

Interlock concrete block paving provides a unique facility for any easy underground work e.g., fire line, cable telephone wire, pipeline repairs, etc., which can be done without losing any time or money as the blocks can be removed and re-laid. Whereas a complete breaking of the pavement is required in the case of R.C.C. flooring, this is not only costly and cumbersome but also consumes a lot of precious time.

7.11   Long term behavior

The efficiency of interlock increases as the number of vehicles passes increases and thus the load distribution to the subgrade improves with time. Whereas the rigid concrete pavement, which is subjected, to stress reversal due to vehicular movement gets damaged due to fatigue.

7.12   Fatigue

As stated above in the case of conventional concrete paving fatigue is an important criterion for designing, whereas in the concrete block, paving being segmental fatigue is not the criterion.

7.13   Low life cycle cost

The initial cost of installation of interlock concrete block paving is cheaper than that of conventional rigid concrete block paving for similar soil and load conditions. Moreover, since the repair and maintenance cost are almost negligible, the interlock concrete block paving is much cheaper than conventional concrete paving, not to mention the valuable operation time which is lost due to time-consuming repair work in case of conventional paving.

The mass production of the interlocking paver block has reduced their price and made it easily affordable. It has less maintenance cost as compared to other paving materials. The concrete block paving is cost-effective as compared to clay pavers or natural stone blocks (granite). Paver block is a long-time investment for your patio and driveway [2].

Conclusion

The paver blocks are used for exterior pavement applications. They are available in different sizes, shapes, colors, designs, thicknesses, textures, and patterns, which lend themselves to the creation of interesting designs in paved areas. It has many advantages over conventional rigid concrete pavement blocks. It is very easily available from manufacturers and suppliers or online as per the choice. The paver block is a durable and long-lasting option for increased home value with the large availability of colors, shapes, and sizes; you can create and also customize patio or driveway design. It gives both the beauty and durability of your pavement at a low cost and low maintenance.

References

[1]   Techno-Economic Feasibility Report on Concrete Hollow and Solid Block, Building Materials and Technology Promotion Council, New Delhi, India, 2010.

[2]   Information on https://gharpedia.com/blog/advantages-of-interlocking-paver-block/

About the Author

Dr. Mohammad Arif Kamal

(Associate Professor)
Architecture Section
Aligarh Muslim University
Aligarh, INDIA
E-mail: architectarif@gmail.com

Dr. Mohammad Arif Kamal is an architect, having around 16 years of Teaching, Research and Professional experience in the field of Architecture, Building Construction Technology and Interior Design. He studied architecture at Aligarh Muslim University, Aligarh from where he received his Bachelor of Architecture (B. Arch.) degree with honors. He was awarded a Government of India, Ministry of Human Resource Development (MHRD) fellowship for pursuing both Master of Architecture (M. Arch.) and Ph.D. degree in Architecture from Indian Institute of Technology (IIT), Roorkee, India. He has designed and implemented many architecture and interior design projects besides his teaching duties both on Design Theory, Construction Systems, Sustainable Architecture, Interior Design as well as Architectural Design Studios.

Dr. Kamal is presently working as an Associate Professor in Aligarh Muslim University, India. He has also worked as Assistant Professor in the Dept. of Architecture at KFUPM, Saudi Arabia for 3 years. He was awarded Junior Faculty Development Grant at KFUPM, Saudi Arabia in 2010. He was associated with an International Architectural Project 'The Mataaf Design Intervention in Haram Mosque', Mecca, Saudi Arabia. His area of research includes Environmental Design, Climate Responsive Architecture, Sustainable Architecture, Building Construction Systems, Building Materials and Traditional Architecture. He has published many Research papers in various International Journals and Conferences as well as many articles in Building Construction Magazines. He has published 3 books and 10 book chapters. He is Editor-in-Chief of 7 International journals related to Architecture, Construction and Civil Engineering. He has edited 5 Special Topic Volume Related to Sustainable Building Materials, published by Trans Tech Publications, Switzerland. He is member of editorial board team/ associate editors of many International journals and conferences and also member of many national and International associations / professional bodies.

www.ingramcontent.com/pod-product-compliance
Lightning Source LLC
Chambersburg PA
CBHW071712210326
41597CB00017B/2452